Physical Sciences

From ancient times, humans have tried to understand the workings of the world around them. The roots of modern physical science go back to the very earliest mechanical devices such as levers and rollers, the mixing of paints and dyes, and the importance of the heavenly bodies in early religious observance and navigation. The physical sciences as we know them today began to emerge as independent academic subjects during the early modern period, in the work of Newton and other 'natural philosophers', and numerous sub-disciplines developed during the centuries that followed. This part of the Cambridge Library Collection is devoted to landmark publications in this area which will be of interest to historians of science concerned with individual scientists, particular discoveries, and advances in scientific method, or with the establishment and development of scientific institutions around the world.

The Works of John Playfair

John Playfair (1748–1819) was a Scottish mathematician and geologist best known for his defence of James Hutton's geological theories. He attended the University of St Andrews, completing his theological studies in 1770. In 1785 he was appointed joint Professor of Mathematics at the University of Edinburgh, and in 1805 he was elected Professor of Natural Philosophy. A Fellow of the Royal Society, he was acquainted with continental scientific developments, and was a prolific writer of scientific articles in the *Transactions of the Royal Society of Edinburgh* and the *Edinburgh Review*. This four-volume edition of his works was published in 1822 and is prefaced by a biography of Playfair. Volume 2 contains the incomplete *Dissertation exhibiting a general view of the progress of mathematical and physical science*, which was included as a supplement to the *Encyclopaedia Britannica*.

Cambridge University Press has long been a pioneer in the reissuing of out-of-print titles from its own backlist, producing digital reprints of books that are still sought after by scholars and students but could not be reprinted economically using traditional technology. The Cambridge Library Collection extends this activity to a wider range of books which are still of importance to researchers and professionals, either for the source material they contain, or as landmarks in the history of their academic discipline.

Drawing from the world-renowned collections in the Cambridge University Library, and guided by the advice of experts in each subject area, Cambridge University Press is using state-of-the-art scanning machines in its own Printing House to capture the content of each book selected for inclusion. The files are processed to give a consistently clear, crisp image, and the books finished to the high quality standard for which the Press is recognised around the world. The latest print-on-demand technology ensures that the books will remain available indefinitely, and that orders for single or multiple copies can quickly be supplied.

The Cambridge Library Collection will bring back to life books of enduring scholarly value (including out-of-copyright works originally issued by other publishers) across a wide range of disciplines in the humanities and social sciences and in science and technology.

The Works of John Playfair

VOLUME 2

JOHN PLAYFAIR
EDITED BY JAMES G. PLAYFAIR

CAMBRIDGE UNIVERSITY PRESS

Cambridge, New York, Melbourne, Madrid, Cape Town,
Singapore, São Paolo, Delhi, Tokyo, Mexico City

Published in the United States of America by Cambridge University Press, New York

www.cambridge.org
Information on this title: www.cambridge.org/9781108029391

This edition first published 1822
This digitally printed version 2011

ISBN 978-1-108-02939-1 Paperback

THE

WORKS

OF

JOHN PLAYFAIR, ESQ.

&c. &c. &c.

Printed by George Ramsay and Company.

THE

WORKS

OF

JOHN PLAYFAIR, ESQ.

LATE

PROFESSOR OF NATURAL PHILOSOPHY IN THE UNIVERSITY OF EDINBURGH,

PRESIDENT OF THE ASTRONOMICAL INSTITUTION OF EDINBURGH,

FELLOW OF THE ROYAL SOCIETY OF LONDON,

SECRETARY OF THE ROYAL SOCIETY OF EDINBURGH,

AND HONORARY MEMBER OF THE ROYAL MEDICAL SOCIETY OF EDINBURGH.

WITH

A MEMOIR OF THE AUTHOR.

VOL. II.

EDINBURGH:

PRINTED FOR ARCHIBALD CONSTABLE & CO. EDINBURGH,

AND HURST, ROBINSON, AND CO. LONDON.

1822.

CONTENTS

OF

VOLUME SECOND.

DISSERTATION:

EXHIBITING A GENERAL VIEW

OF THE PROGRESS OF

MATHEMATICAL AND PHYSICAL SCIENCE,

SINCE THE REVIVAL OF LETTERS IN EUROPE.

ADVERTISEMENT.

This Dissertation was published in the Second and Fourth Volumes of the Supplement to the Encyclopædia Britannica, in which the following account of it is given:

"Mr Playfair's Dissertation was intended to furnish an Historical Sketch of the principal Discoveries and Improvements in Science, from the Revival of Letters to the beginning of the present Century; and, in that portion of it which is prefixed to the *Second* Volume of this work, the history is brought down to the period marked by the commencement of Newton's discoveries. The remaining half was to have completed the design, in *three* parts or subdivisions; the First, comprehending the period of Newton and Leibnitz; the Second, that of Euler and D'Alembert; and the Third, that of Lagrange and Laplace.

"Mr Playfair was proceeding, with his accustomed diligence and ardour, in the execution of this interesting and congenial task, when he was seized with the illness of which he died. The *first* subdivision of this part of his plan, which embraces a view of the advances made in the most remarkable period of the history of Science, was happily completed, and the printing finished, while he was yet able to correct the press. It is now given to the Public, under the painful impression that it must too probably be considered as a *Fragment;* for the Editor fears, that the materials collected for the completion of the Dissertation, though containing the results of much elaborate inquiry, and profound reflection, cannot be put into a shape that would justify their publication as a work of Professor Playfair.

"*Edinburgh, December* 1819."

The opinion entertained by the Editor of the Supplement having since been fully confirmed by a careful examination of the manuscript, the Dissertation is here given as it was originally published in that work.

DISSERTATION.

PART I.

In conformity to the plan which has been traced and executed with so much ability in the First Dissertation,* I am now to present the reader with an historical sketch of the principal discoveries made in Natural Philosophy, from the revival of letters down to the present time. In entering on this task, and on looking at the instructive but formidable model already set before me, I should experience no small solicitude, did I not trust that the subject of which I am to speak, in order to be interesting, needs only to be treated with clearness and precision. These two requisites I will endeavour to keep steadily in view.

In the order which I am to follow, I shall be guided solely by a regard to the subserviency of one science to the progress of another, and to the consequent priority of the former in the order of

* The Dissertation here alluded to is the work of Mr Stewart, containing the *History of Metaphysical, Ethical, and Political Philosophy*.—E.

regular study. For this reason, the history of the pure Mathematics will be first considered, as that science has been one of the two principal instruments applied by the moderns to the advancement of natural knowledge. The other instrument is Experience; and, therefore, the principles of the inductive method, or of the branch of Logic which teaches the application of experiment and observation to the interpretation of nature, must be the second object of inquiry; and in this article I shall give an account of Bacon's Philosophy, as applied to Physical investigation. After these two sections, which may in some measure be considered as introductory, I am to treat of Natural Philosophy, under the divisions of Mechanics, Astronomy, and Optics. Under the general denomination of Mechanics I include the Theory of Motion, as applied not only to solids, but to fluids, both incompressible and elastic. Optics I have placed after Astronomy, because the discoveries in Mechanics have much less affected the progress of the former of these sciences than of the latter. To these will succeed a sixth division, containing the laws of the three unknown substances, if, indeed, they may be called substances,—Heat, Electricity, and Magnetism. These, though very different, agree in some general characters. They permeate all substances, though not with the same facility; and, if other bodies had been formed in the same manner with them, the idea of impenetrability would never have been suggested to the mind. They seem to receive motion, without taking any away from the body which communicates it; so that they can hardly be considered as inert. Two of them, Heat and Electricity, are perceived by the sense of touch; but the impression which they make does not convey any idea of resistance. The third is not perceived by touch; and, therefore, all the three might be denominated impalpable substances. If they have any gravity, it cannot be appreciated; and, for these reasons, had it not too paradoxical an appearance, we might class them together as *material* but *incorporeal* substances. We know, indeed, nothing of them but as powers, transferable from one body to another; and it is in consequence of this last circumstance alone that they are entitled to the name of substances.

Though the general design of this historical sketch extends from the revival of letters to the beginning of the nineteenth century, I shall, in the present Part, confine myself entirely, as has been done in the first Discourse, to the period preceding the end of the seventeenth century, or, more precisely, to that preceding the invention of the fluxionary calculus, and the discovery of the principle of gravitation;—one of the most remarkable epochas, without doubt, in the history of human knowledge.

Section I.

MATHEMATICS.

1. Geometry.

The great inheritance of mathematical knowledge which the ancients bequeathed to posterity could not, on the revival of learning, be immediately taken possession of, nor could even its existence be discovered, but by degrees. Though the study of the Mathematics had never been entirely abandoned, it had been reduced to matters of very simple and easy comprehension, such as were merely subservient to practice. There had been men who could compute the area of a triangle, draw a meridian line, or even construct a sun-dial, in the worst of times; but between such skill, and the capacity to understand or the taste to relish, the demonstrations of Euclid, Apollonius, or Archimedes, there was a great interval, and many difficulties were to be overcome, for which much time, and much subsidiary knowledge, were necessary. The repositories of the ancient treasures were to be opened, and made accessible; the knowledge of the languages was to be acquired; the manuscripts were to be decyphered; and the skill of the grammarian and the critic were to precede, in a certain degree, that of the geometrician or the astronomer. The obligations which we have to those who undertook this laborious and irksome task, and who rescued the ancient books from the prisons to which ignorance and barbarism had condemned them, and from the final destruction by which they must soon have been overtaken, are such as we can never sufficiently acknowledge; and, indeed, we shall never know even the names of many of the benefactors to whom our thanks are due. In the midst of the wars, the confusion, and bloodshed, which overwhelmed Europe during the middle ages, the religious houses and monasteries afforded to the remains of ancient learning an asylum, which a salutary prejudice forced even the most lawless to respect; and the authors who have given the best account of the revival of letters, agree that it is in a great measure to those establishments that we owe the safety of the books which have kept alive the scientific and literary attainments of Greece and Rome.

The study of the remains of antiquity gradually produced men of taste and intelligence, who were able to correct the faults of the manuscripts they copied, and to explain the difficulties of the authors they translated. Such were Purbach, Regiomontanus, Commandine, Maurolycus, and many others. By their means, the writings of Euclid, Archimedes, Apollonius, Ptolemy, and Pappus, became known and accessible to men of science. Arabia contributed its share towards this great renovation, and from the language of that country was derived the knowledge of many Greek books, of the originals of which, some were not found till long afterwards, and others have never yet been discovered.

In nothing, perhaps, is the inventive and elegant genius of the Greeks better exemplified than in their geometry. The elementary truths of that science were connected by Euclid into one great chain, beginning from the axioms, and extending to the properties of the five regular solids; the whole digested into such admirable order, and explained with such clearness and precision, that no similar work of superior excellence has appeared, even in the present advanced state of mathematical science.

Archimedes had assailed the more difficult problems of geometry, and by means of the method of Exhaustions, had demonstrated many curious and important theorems, with regard to the lengths and areas of curves, and the contents of solids. The same great geometer had given a beginning to physico-mathematical science, by investigating several propositions, and resolving several problems in Mechanics and Hydrostatics.

Apollonius had treated of the Conic Sections,—the Curves which, after the circle, are the most simple and important in geometry; and, by his elaborate and profound researches, had laid the foundation of discoveries which were to illustrate very distant ages.

Another great invention, the Geometrical Analysis, ascribed very generally to the Platonic school, but most successfully cultivated by the geometer just named, is one of the most ingenious and beautiful contrivances in the mathematics. It is a method of discovering truth by reasoning concerning things unknown, or propositions merely supposed, as if the one were given, or the other were really true. A quantity that is unknown, is only to be found from the relations which it bears to quantities that are known. By reasoning on these relations, we come at last to some one so simple, that the thing sought is thereby determined. By this analytical process, therefore, the thing required is discovered, and we are at the same time put in possession of an instrument by which new truths may be found out, and which, when skill in using it has been acquired by practice, may be applied to an unlimited extent.

A similar process enables us to discover the demonstrations of propositions, supposed to be true, or, if not true, to discover that they are false.

This method, to the consideration of which we shall again have an opportunity of returning, was perhaps the most valuable part of the ancient mathematics, inasmuch as a method of discovering truth is more valuable than the truths it has already discovered. Unfortunately, however, the fragments containing this precious remnant had suffered more from the injuries of time than almost any other.

In the fifteenth century, Regiomontanus, already mentioned, is the mathematician who holds the highest rank. To him we owe many translations and commentaries, together with several original and valuable works of his own. Trigonometry, which had never been known to the Greeks as a separate science, and which took that form in Arabia, advanced, in the hands of Regiomontanus, to a great degree of perfection, and approached very near to the condition which it has attained at the present day. He also introduced the use of decimal fractions into arithmetic, and thereby gave to that scale its full extent, and to numerical computation the utmost degree of simplicity and enlargement which it seems capable of attaining.

This eminent man was cut off in the prime of life; and his untimely death, says Mr Smith, amidst innumerable projects for the advancement of science, is even at this day a matter of regret.* He was buried in the Pantheon at Rome; and the honours paid to him at his death prove that science had now become a distinction which the great were disposed to recognize.

Werner, who lived in the end of this century, is the first among the moderns who appears to have been acquainted with the geometrical analysis. His writings are very rare, and I have never had an opportunity of examining them. What I here assert is on the authority of Montucla, whose judgment in this matter may be safely relied on, as he has shown, by many instances, that he was well acquainted with the nature of the analysis referred to. It is not a little remarkable that Werner should have understood this subject, when we find many eminent mathematicians, long after his time, entirely unacquainted with it, and continually expressing their astonishment how the ancient geometers found out those simple and elegant constructions and demonstrations, of which they have given so many examples. In the days of Werner, there was no ancient book known except the *Data* of Euclid, from which any information concerning the geometrical analysis could be collected; and it is highly to his credit, that, without any other help,

---

* History of Astronomy, p. 90. Regiomontanus was born in 1436, and died in 1476.

he should have come to the knowledge of a method not a little recondite in its principles, and among the finest inventions either of ancient or of modern science. Werner resolved, by means of it, Archimedes's problem of cutting a sphere into two segments, having a given ratio to one another. He proposed also to translate, from the Arabic, the work of Apollonius, entitled *Sectio Rationis*, rightly judging it to be an elementary work in that analysis, and to come next after the *Data* of Euclid. *

Benedetto, an Italian mathematician, appears also to have been very early acquainted with the principles of the same ingenious method, as he published a book on the geometrical analysis at Turin in 1585.

Maurolycus of Messina flourished in the middle of the sixteenth century, and is justly regarded as the first geometer of that age. Beside furnishing many valuable translations and commentaries, he wrote a treatise on the conic sections, which is highly esteemed. He endeavoured also to restore the fifth book of the conics of Apollonius, in which that geometer treated of the *maxima* and *minima* of the conic sections. His writings all indicate a man of clear conceptions, and of a strong understanding; though he is taxed with having dealt in astrological prediction.

* See Montucla, Vol. I. p. 470.

In the early part of the seventeenth century, Cavalieri was particularly distinguished, and made an advance in the higher geometry, which occupies the middle place between the discoveries of Archimedes and those of Newton.

For the purpose of determining the lengths and areas of curves, and the contents of solids contained within curve superficies, the ancients had invented a method, to which the name of Exhaustions has been given; and in nothing, perhaps, have they more displayed their powers of mathematical invention.

Whenever it is required to measure the space bounded by curve lines, the length of a curve, or the solid contained within a curve superficies, the investigation does not fall within the range of elementary geometry. Rectilineal figures are compared, on the principle of superposition, by help of the notion of equality which is derived from the coincidence of magnitudes both similar and equal. Two rectangles of equal bases and equal altitudes are held to be equal, because they can perfectly coincide. A rectangle and an oblique angled parallelogram, having equal bases and altitudes, are shown to be equal, because the same triangle, taken from the rectangle on one side, and added to it on the other, converts it into the parallelogram; and thus two magnitudes which are not similar, are shown to have equal areas. In like manner, if a triangle and a parallelogram have the same base and altitude, the triangle is shown to be half the parallelogram; because, if to the triangle there be added another, similar and equal to itself, but in the reverse position, the two together will compose a parallelogram, having the same base and altitude with the given triangle. The same is true of the comparison of all other rectilineal figures; and if the reasoning be carefully analyzed, it will always be found to be reducible to the primitive and original idea of equality, derived from things that coincide or occupy the same space; that is to say, the areas which are proved equal are always such as, by the addition or subtraction of equal and similar parts, may be rendered capable of coinciding with one another.

This principle, which is quite general with respect to rectilineal figures, must fail, when we would compare curvilineal and rectilineal spaces with one another, and make the latter serve as measures of the former, because no addition or subtraction of rectilineal figures can ever produce a figure which is curvilineal. It is possible, indeed, to combine curvilineal figures, so as to produce one that is rectilineal; but this principle is of very limited extent; it led to the quadrature of the *lunulæ* of Hippocrates, but has hardly furnished any other result which can be considered as valuable in science.

In the difficulty to which geometers were thus reduced, it might occur, that, by inscribing a rectilineal figure within a curve, and circumscribing another round it, two limits could be obtained, one greater and the other less than the area required. It was also evident, that, by increasing the number, and diminishing the sides of those figures, the two limits might be brought continually nearer to one another, and of course nearer to the curvilineal area, which was always intermediate between them. In prosecuting this sort of approximation, a result was at length found out, which must have occasioned no less surprise than delight to the mathematician who first encountered it. The result I mean is, that, when the series of inscribed figures was continually increased, by multiplying the number of the sides, and diminishing their size, there was an assignable *rectilineal* area, to which they continually approached, so as to come nearer it than any difference that could be supposed. The same limit would also be observed to belong to the circumscribed figures, and therefore it could be no other than the curvilineal area required.

It appears to have been to Archimedes that a truth of this sort first occurred, when he found that two-thirds of the rectangle, under the ordinate and abscissa of a parabola, was a limit always greater than the inscribed rectilineal figure, and less than the circumscribed. In some other curves, a similar conclusion was found, and Archimedes contrived to show that it was impossible to suppose that the area of the curve could differ from the said limit, without admitting that the circumscribed figure might become less, or the inscribed figure greater than the curve itself. The method of Exhaustions was the name given to the indirect demonstrations thus formed. Though few things more ingenious than this method have been devised, and though nothing could be more conclusive than the demonstrations resulting from it, yet it laboured under two very considerable defects. In the first place, the process by which the demonstration was obtained was long and difficult; and, in the second place, it was indirect, giving no insight into the principle on which the investigation was founded. Of consequence, it did not enable one to find out similar demonstrations, nor increase one's power of making more discoveries of the same kind. It was a demonstration purely synthetical, and required, as all indirect reasoning must do, that the conclusion should be known before the reasoning is begun. A more compendious, and a more analytical method, was therefore much to be wished for, and was an improvement, which, at a moment when the field of mathematical science was enlarging so fast, seemed particularly to be required.

Cavalieri, born at Milan in the year 1598, is the person by whom this great improvement was made.

The principle on which he proceeded was, that areas may be considered as made up of an infinite number of parallel lines; solids of an infinite number of parallel planes; and even lines themselves, whether curve or straight, of an infinite number of points. The cubature of a solid being thus reduced to the summation of a series of planes, and the quadrature of a curve to the summation of a series of ordinates, each of the investigations was reduced to something more simple. It added to this simplicity not a little, that the sums of series are often more easily found, when the number of terms is infinitely great, than when it is finite, and actually assigned.

It appears that a tract on stereometry, written by Kepler, whose name will hereafter be often mentioned, first led Cavalieri to take this view of geometrical magnitudes. In that tract, which was published in 1615, the measurement of many solids was proposed, which had not before fallen under the consideration of mathematicians. Such, for example, was that of the solids generated by the revolution of a curve, not about its axis, but about any line whatsoever. Solids of that kind, on account of their affinity with the figure of casks, and vessels actually employed for containing liquids, appeared to Kepler to offer both curious and useful subjects of investigation. There were no less than eighty-four such solids, which he proposed for the consideration of mathematicians. He was, however, himself unequal to the task of resolving any but a small number of the simplest of these problems. In these solutions, he was bold enough to introduce into geometry, for the first time, the idea of infinitely great and infinitely small quantities, and by this apparent departure from the rigour of the science, he rendered it in fact a most essential service. Kepler conceived a circle to be composed of an infinite number of triangles, having their common vertex in the centre of the circle, and their infinitely small bases in the circumference. It is to be remarked, that Galileo had also introduced the notion of infinitely small quantities, in his first dialogue, De Mechanica, where he treats of a cylinder cut out of a hemisphere ; and he has done the same in treating of the acceleration of falling bodies. Cavalieri was the friend and disciple of Galileo, but much more profound in the mathematics. In his hands the idea took a more regular and systematic form, and was explained in his work on indivisibles, published in 1635.

The rule for summing an infinite series of terms in arithmetical progression had been long known, and the application of it to find the area of a triangle, according to the method of indivisibles, was a matter of no difficulty. The next step was, supposing a series of lines in arithmetical progression, and squares to be described on each of them, to find what ratio the sum of all these squares bears to the greatest square, taken as often as there are terms in the progression. Cavalieri showed, that when the number of terms is infinitely great, the first of these sums is just one-third of the second. This evidently led to the cubature of many solids.

Proceeding one step farther, he sought for the sum of the cubes of the same lines, and found it to be one-fourth of the greatest, taken as often as there are terms; and, continuing this investigation, he was able to assign the sum of the nth powers of a series in arithmetical progression, supposing always the difference of the terms to be infinitely small, and their number to be infinitely great. The number of curious results obtained from these investigations may be easily conceived. It gave, over geometrical problems of the higher class, the same power which the integral calculus, or the inverse method of fluxions does, in the case when the exponent of the variable quantity is an integer. The method of indivisibles, however, was not without difficulties, and could not but be liable to objection, with those accustomed to the rigorous exactness of the ancient geometry. In strictness, lines, however multiplied, can never make an area, or any thing but a line; nor can areas, however they may be added together, compose a solid, or any thing but an area. This is certainly true, and yet the conclusions of Cavalieri, deduced on a contrary supposition, are true also. This happened, because, though the suppositions that a certain series of lines, infinite in number, and contiguous to one another, may compose a certain area, and that another series may compose another area, are neither of them true; yet is it strictly true, that the one of these areas must have to the other the same ratio which the sum of the one series of lines has to the sum of the other series. Thus, it is the ratios of the areas, and not the areas absolutely considered, which are determined by the reasonings of Cavalieri; and that this determination of their ratios is quite accurate, can very readily be demonstrated by the method of exhaustions.

The method of indivisibles, from the great facility with which it could be managed, furnished a most ready method of ascertaining the ratios of areas and solids to one another, and, therefore, scarcely seems to deserve the epithet which Newton himself bestows upon it, of involving in its conceptions something harsh, (*durum,*) and not easy to be admitted. It was the doctrine of infinitely small quantities carried to the extreme, and gave at once the result of an infinite series of successive approximations. Nothing, perhaps, more ingenious, and certainly nothing more happy, ever was contrived, than to arrive at the conclusion of all these approximations, without going through the approximations themselves. This is the purpose served by introducing into mathematics the consideration of quantities infinitely small in size, and infinitely great in number; ideas which, however inaccurate they may seem, yet, when carefully and analogically reasoned upon, have never led into error.

Geometry owes to Cavalieri, not only the general method just described, but many particular theorems, which that method was the instrument of discovering. Among these is the very remarkable proposition, *that as four right angles, to the excess of the three angles of any spherical triangle, above two right angles, so is the superficies of the hemisphere to the area of the triangle.* At that time, however, science was advancing so fast, and the human mind was everywhere expanding itself with so much energy, that the same discovery was likely to be made by more individuals than one at the same time. It was not known in Italy in 1632, when this determination of the area of a spherical triangle was given by Cavalieri, that it had been published three years before by Albert Girard, a mathematician of the Low Countries, of whose inventive powers we shall soon have more occasion to speak.

The Cycloid afforded a number of problems, well calculated to exercise the proficients in the geometry of indivisibles, or of infinites. It is the curve described by a point in the circumference of a circle, while the circle itself rolls in a straight line along a plane. It is not quite certain when this curve, so remarkable for its curious properties, and for the place which it occupies in the history of geometry, first drew the attention of mathematicians. In the year 1639, Galileo informed his friend Torricelli, that, forty years before that time, he had thought of this curve, on account of its shape, and the graceful form it would give to arches in architecture. The same philosopher had endeavoured to find the area of the cycloid; but though he was one of those who first introduced the consideration of infinites into geometry, he was not expert enough in the use of that doctrine, to be able to resolve this problem. It is still more extraordinary, that the same problem proved too difficult for Cavalieri, though he certainly was in complete possession of the principles by which it was to be resolved. It is, however, not easy to determine whether it be to Torricelli, the scholar of Cavalieri, and his successor in genius and talents, or to Roberval, a French mathematician of the same period, and a man also of great originality and invention, that science is indebted for the first quadrature of the cycloid, or the proof that its area is three times that of its generating circle. Both these mathematicians laid claim to it. The French and Italians each took the part of their own countryman; and in their zeal have so perplexed the question, that it is hard to say on which side the truth is to be found.

Torricelli, however, was a man of a mild, amiable, and candid disposition; Roberval of a temper irritable, violent, and envious; so that, in as far as the testimony of the individuals themselves is concerned, there is no doubt which ought to preponderate. They had both the skill and talent which fitted them for this, or even for more difficult researches.

The other properties of this curve, those that respect its tangents, its length, its curvature, &c. exercised the ingenuity, not only of the geometers just mentioned, but of Wren, Wallis, Huygens, and, even after the invention of the integral calculus, of Newton, Leibnitz, and Bernoulli.

Roberval also improved the method of quadratures invented by Cavalieri, and extended his solutions to the case, when the powers of the terms in the arithmetical progression of which the sum was to be found were fractional; and Wallis added the case when they were negative. Fermat, who, in his inventive resources, as well as in the correctness of his mathematical taste,* yielded to none of his contemporaries, applied the consideration of infinitely small quantities to determine the *maxima* and *minima* of the ordinates of curves, as also their

* He also was very skilful in the geometric analysis, and seems to have more thoroughly imbibed the spirit of that ingenious invention than any of the moderns before Halley.

tangents. Barrow, somewhat later, did the same in England. Afterwards the geometry of infinites fell into the hands of Leibnitz and Newton, and acquired that new character which marks so distinguished an era in the mathematical sciences.

2. ALGEBRA.

It was not from Greece alone that the light proceeded which dispelled the darkness of the middle ages; for, with the first dawn of that light, a mathematical science, of a name and character unknown to the geometers of antiquity, was received in Europe from Arabia. As early as the beginning of the thirteenth century, Leonardo, a merchant of Pisa, having made frequent visits to the East, in the course of commercial adventure, returned to Italy, enriched by the traffic, and instructed by the science of those countries. He brought with him the knowledge of *Algebra*; and a late writer quotes a manuscript of his, bearing the date of 1202, and another that of 1228.* The importation of Algebra into Europe, is thus carried back nearly 200 years farther than has generally been supposed, for Leonardo has been represented as flourishing in the

* M. Cossali of Pisa, in a Tract on the Origin of Algebra, 1797.

end of the fourteenth century, instead of the very beginning of the thirteenth. It appears by an extract from his manuscript, published by the above author, that his knowledge of Algebra extended as far as quadratic equations. The language was very imperfect, corresponding to the infancy of the science; the quantities and the operations being expressed in words, with the help only of a few abbreviations. The rule for resolving quadratics by completing the square, is demonstrated geometrically.

Though Algebra was brought into Europe from Arabia, it is by no means certain that this last is its native country. There is, indeed, reason to think that its invention must be sought for much farther to the East, and probably not nearer than Indostan. We are assured by the Arabian writers, that Mahomet Ben Musa of Chorasan, distinguished for his mathematical knowledge, travelled, about the year 959, into India, for the purpose of receiving farther instruction in the science which he cultivated. It is likewise certain, that some books, which have lately been brought from India into this country, treat of Algebra in a manner that has every appearance of originality, or at least of being derived from no source with which we are at all acquainted.

Before the time of Leonardo of Pisa, an important acquisition, also from the East, had greatly improved the science of arithmetic. This was the use of the Arabic notation, and the contrivance of making the same character change its signification, according to a fixed rule, when it changed its position, being increased tenfold for every place that it advanced towards the left. The knowledge of this simple but refined artifice was learned from the Moors by Gerbert, a monk of the Low Countries, in the tenth century, and by him made known in Europe. Gerbert was afterwards Pope, by the name of Silvester the Second; but from that high dignity derived much less glory than from having instructed his countrymen in the decimal notation.

The writings of Leonardo, above mentioned, have remained in manuscript; and the first printed book in Algebra is that of Lucas de Burgo, a Franciscan, who, towards the end of the fifteenth century, travelled, like Leonardo, into the East, and was there instructed in the principles of Algebra. The characters employed in his work, as in those of Leonardo, are mere abbreviations of words. The letters *p* and *m* denote *plus* and *minus*; and the rule is laid down, that, in multiplication, *plus* into *minus* gives *minus*, but *minus* into *minus* gives *plus*. Thus the first appearance of Algebra is merely that of a system of short-hand writing, or an abbreviation of common language, applied to the solution of arithmetical problems.

It was a contrivance merely to save trouble ; and yet to this contrivance we are indebted for the most philosophical and refined art which men have yet employed for the expression of their thoughts. This scientific language, therefore, like those in common use, has grown up slowly, from a very weak and imperfect state, till it has reached the condition in which it is now found.

Though in all this the moderns received none of their information from the Greeks, yet a work in the Greek language, treating of arithmetical questions, in a manner that may be accounted algebraic, was discovered in the course of the next century, and given to the world, in a Latin translation, by Xylander, in 1575. This is the work of Diophantus of Alexandria, who had composed thirteen books of Arithmetical Questions, and is supposed to have flourished about 150 years after the Christian era. The questions he resolves are often of considerable difficulty ; and a great deal of address is displayed in stating them, so as to bring out equations of such a form, as to involve only one power of the unknown quantity. The expression is that of common language, abbreviated and assisted by a few symbols. The investigations do not extend beyond quadratic equations ; they are, however, extremely ingenious, and prove the author to have been a man of talent, though the instrument he worked with was weak and imperfect.

The name of Cardan is famous in the history of Algebra. He was born at Milan in 1501, and was a man in whose character good and ill, strength and weakness, were mixed up in singular profusion. With great talents and industry, he was capricious, insincere, and vain-glorious to excess. Though a man of real science, he professed divination, and was such a believer in the influence of the stars, that he died to accomplish an astrological prediction. He remains, accordingly, a melancholy proof, that there is no folly or weakness too great to be united to high intellectual attainments.

Before his time, very little advance had been made in the solution of any equations higher than the second degree; except that, as we are told, about the year 1508, Scipio Ferreo, Professor of Mathematics at Bologna, had found out a rule for resolving one of the cases of cubic equations, which, however, he concealed, or communicated only to a few of his scholars. One of these, Florido, on the strength of the secret he possessed, agreeably to a practice then common among mathematicians, challenged Tartalea of Brescia, to contend with him in the solution of algebraic problems. Florido had at first the advantage; but Tartalea, being a man of ingenuity, soon discovered his rule, and also another much more general, in consequence of which, he came off at last victorious. By the report of this victory, the curiosity of Cardan was strongly excited; for, though he was himself much versed in the mathematics, he had not been able to discover a method of resolving equations higher than the second degree. By the most earnest and importunate solicitation, he wrung from Tartalea the secret of his rules, but not till he had bound himself, by promises and oaths, never to divulge them. Tartalea did not communicate the demonstrations, which, however, Cardan soon found out, and extended, in a very ingenious and systematic manner, to all cubic equations whatsoever. Thus possessed of an important discovery, which was at least in a great part his own, he soon forgot his promises to Tartalea, and published the whole in 1545, not concealing, however, what he owed to the latter. Though a proceeding, so directly contrary to an express stipulation, cannot be defended, one does not much regret the disappointment of any man who would make a mystery of knowledge, or keep his discoveries a secret, for purposes merely selfish.

Thus was first published the rule which still bears the name of Cardan, and which, at this day, marks a point in the progress of algebraic investigation, which all the efforts of succeeding analysts have hardly been able to go beyond. As to the general doctrine of equations, it appears that Cardan was acquainted both with the negative and positive roots, the former of which he called by the name of false roots. He also knew that the number of positive, or, as he called them, true roots, is equal to the number of the changes of the signs of the terms; and that the coefficient of the second term is the difference between the sum of the true and the false roots. He also had perceived the difficulty of that case of cubic equations, which cannot be reduced to his own rule. He was not able to overcome the difficulty, but showed how, in all cases, an approximation to the roots might be obtained.

There is the more merit in these discoveries, that the language of Algebra still remained very imperfect, and consisted merely of abbreviations of words. Mathematicians were then in the practice of putting their rules into verse. Cardan has given his a poetical dress, in which, as may be supposed, they are very awkward and obscure; for whatever assistance in this way is given to the memory, must be entirely at the expence of the understanding. It is, at the same time, a proof that the language of Algebra was very imperfect. Nobody now thinks of translating an algebraic formula into verse; because, if one has acquired any familiarity with the language of the science, the formula will be more easily remembered than any thing that can be substituted in its room.

Italy was not the only country into which the algebraic analysis had by this time found its way;

4

in Germany it had also made considerable progress, and Stiphelius, in a book of Algebra, published at Nuremberg in 1544, employed the same numeral exponents of powers, both positive and negative, which we now use, as far as integer numbers are concerned; but he did not carry the solution of equations farther than the second degree. He introduced the same characters for *plus* and *minus* which are at present employed.

Robert Recorde, an English mathematician, published about this time, or a few years later, the first English treatise on Algebra, and he there introduced the same sign of equality which is now in use.

The properties of algebraic equations were discovered, however, very slowly. Pelitarius, a French mathematician, in a treatise which bears the date of 1558, is the first who observed that the root of an equation is a divisor of the last term; and he remarked also this curious property of numbers, that the sum of the cubes of the natural numbers is the square of the sum of the numbers themselves.

The knowledge of the solution of cubic equations was still confined to Italy. Bombelli, a mathematician of that country, gave a regular treatise on Algebra, and considered, with very particular attention, the irreducible case of Cardan's rule. He was the first who made the remark, that the problems belonging to that case can always be resolved by the trisection of an arch. *

Vieta was a very learned man, and an excellent mathematician, remarkable both for industry and invention. He was the first who employed letters to denote the known as well as the unknown quantities, so that it was with him that the language of algebra first became capable of expressing general truths, and attained to that extension which has since rendered it such a powerful instrument of investigation. He has also given new demonstrations of the rule for resolving cubic, and even bi-

* A passage in Bombelli's book, relative to the Algebra of India, has become more interesting, from the information concerning the science of that country, which has reached Europe within the last twenty years. He tells us, that he had seen in the Vatican library, a manuscript of a certain Diophantus, a Greek author, which he admired so much, that he had formed the design of translating it. He adds, that in this manuscript he had found the Indian authors often quoted; from which it appeared, that Algebra was known to the Indians before it was known to the Arabians. Nothing, however, of all this is to be found in the work of Diophantus, which was published about three years after the time when Bombelli wrote. As it is, at the same time, impossible that he could be so much mistaken about a manuscript which he had particularly examined, this passage remains a mystery, which those who are curious about the ancient history of science would be very glad to have unravelled. See Hutton's History of Algebra.

quadratic equations. He also discovered the relation between the roots of an equation of any degree, and the coefficients of its terms, though only in the case where none of the terms are wanting, and where all the roots are real or positive. It is, indeed, extremely curious to remark, how gradually the truths of this sort came in sight. This proposition belonged to a general truth, the greater part of which remained yet to be discovered. Vieta's treatises were originally published about the year 1600, and were afterwards collected into one volume by Schooten, in 1646.

In speaking of this illustrious man, Vieta, we must not omit his improvements in trigonometry, and still less his treatise on angular sections, which was a most important application of Algebra to investigate the theorems, and resolve the problems of geometry. He also restored some of the books of Apollonius, in a manner highly creditable to his own ingenuity, but not perfectly in the taste of the Greek geometry; because, though the constructions are elegant, the demonstrations are all synthetical.

About the same period, Algebra became greatly indebted to Albert Girard, a Flemish mathematician, whose principal work, Invention Nouvelle en Algebre, was printed in 1669. This ingenious author perceived a greater extent, but not yet the whole of the truth, partially discovered by Vieta, viz. the successive formation of the coefficients of an equation from the sum of the roots; the sum of their products taken two and two; the same taken three and three, &c. whether the roots be positive or negative. He appears also to have been the first who understood the use of negative roots in the solution of geometrical problems, and is the author of the figurative expression, which gives to negative quantities the name of *quantities less than nothing*; a phrase that has been severely censured by those who forget that there are correct ideas, which correct language can hardly be made to express. The same mathematician conceived the notion of imaginary roots, and showed that the number of the roots of an equation could not exceed the exponent of the highest power of the unknown quantity. He was also in possession of the very refined and difficult rule, which forms the sums of the powers of the roots of an equation from the coefficients of its terms. This is the greatest list of discoveries which the history of any algebraist could yet furnish.

The person next in order, as an inventor in Algebra, is Thomas Harriot, an English mathematician, whose book, Artis Analyticæ Praxis, was published after his death, in 1631. This book contains the genesis of all equations, by the continued multiplication of simple equations; that is to say, it explains the truth in its full extent, to which Vieta and Girard had been approximating. By

Harriot also, the method of extracting the roots of equations was greatly improved; the smaller letters of the alphabet, instead of the capital letters employed by Vieta, were introduced; and by this improvement, trifling, indeed, compared with the rest, the form and exterior of algebraic expression were brought nearer to those which are now in use.

I have been the more careful to note very particularly the degrees by which the properties of equations were thus unfolded, because I think it forms an instance hardly paralleled in science, where a succession of able men, without going wrong, advanced, nevertheless, so slowly in the discovery of a truth which, when known, does not seem to be of a very hidden and abstruse nature. Their slow progress arose from this, that they worked with an instrument, the use of which they did not fully comprehend, and employed a language which expressed more than they were prepared to understand;—a language which, under the notion, first of negative and then of imaginary quantities, seemed to involve such mysteries as the accuracy of mathematical science must necessarily refuse to admit.

The distinguished author of whom I have just been speaking was born at Oxford in 1560. He was employed in the second expedition sent out by Sir Walter Raleigh to Virginia, and on his return published an account of that country. He afterwards devoted himself entirely to the study of the mathematics; and it appears from some of his manuscripts, lately discovered, that he observed the spots of the sun as early as December 1610, not more than a month later than Galileo. He also made observations on Jupiter's satellites, and on the comets of 1607, and of 1618.*

The succession of discoveries, above related, brought the algebraic analysis, abstractly considered, into a state of perfection, little short of that which it has attained at the present moment. It was thus prepared for the step which was about to be taken by Descartes, and which forms one of the most important epochas in the history of the mathematical sciences. This was the application of the algebraic analysis, to define the nature, and investigate the properties, of curve lines, and, consequently, to represent the notion of variable quantity. It is often said, that Descartes was the first who applied algebra to geometry; but this is inaccurate; for such applications had been made be-

---

* The manuscripts which contain these observations, and probably many other things of great interest, are preserved in the collection of the Earl of Egremont, having come into the possession of his family from Henry Percy Earl of Northumberland, a most liberal patron of science, with whom Harriot appears to have chiefly lived after his return from Virginia.

fore, particularly by Vieta, in his treatise on angular sections. The invention just mentioned is the undisputed property of Descartes, and opened up vast fields of discovery for those who were to come after him.

The work in which this was contained is a tract of no more than 106 quarto pages; and there is probably no book of the same size which has conferred so much and so just celebrity on its author. It was first published in 1637.

In the first of the three books into which the tract just mentioned is divided, the author begins with the consideration of such geometrical problems as may be resolved by circles and straight lines; and explains the method of constructing algebraic formulas, or of translating a truth from the language of algebra into that of geometry. He then proceeds to the consideration of the problem, known among the ancients by the name of the *locus ad quatuor rectas*, and treated of by Apollonius and Pappus. The algebraic analysis afforded a method of resolving this problem in its full extent; and the consideration of it is again resumed in the second book. The thing required is, to find the locus of a point, from which, if perpendiculars be drawn to four lines given in position, a given function of these perpendiculars, in which the variable quantities are only of two dimensions, shall be always of the same magnitude.* Descartes shows the locus, on this hypothesis, to be always a conic section; and he distinguishes the cases in which it is a circle, an ellipsis, a parabola, or a hyperbola. It was an instance of the most extensive investigation which had yet been undertaken in geometry, though, to render it a complete solution of the problem, much more detail was doubtless necessary. The investigation is extended to the cases where the function, which remains the same, is of three, four, or five dimensions, and where the locus is a line of a higher order, though it may, in certain circumstances, become a conic section. The lines given in position may be more than four, or than any given number; and the lines drawn to them may either be perpendiculars, or lines making given angles with them. The same analysis applies to all the cases; and this problem, therefore, afforded an excellent example of the use of algebra in the investigation of geometrical propositions. The author takes notice of the unwillingness of the ancients to transfer the language of arithmetic into

* It will easily be perceived, that the word *function* is not contained in the original enunciation of the problem. It is a term but lately introduced into mathematical language, and affords here, as on many other occasions, a more general and more concise expression than could be otherwise obtained.

geometry, so that they were forced to have recourse to very circuitous methods of expressing those relations of quantity in which powers beyond the third are introduced. Indeed, to deliver investigation from those modes of expression which involve the composition of ratios, and to substitute in their room the multiplication of the numerical measures, is of itself a very great advantage, arising from the introduction of algebra into geometry.

In this book also, an ingenious method of drawing tangents to curves is proposed by Descartes, as following from his general principles, and it is an invention with which he appears to have been particularly pleased. He says, "Nec verebor dicere problema hoc non modo eorum, quæ scio, utilissimum et generalissimum esse, sed etiam eorum quæ in geometria scire unquam desideraverim." * This passage is not a little characteristic of Descartes, who was very much disposed to think well of what he had done himself, and even to suppose that it could not easily be rendered more perfect. The truth, however, is, that his method of drawing tangents is extremely operose, and is one of those hasty views which, though ingenious and even profound, require to be vastly simplified, before they can be reduced to practice. Fermat, the rival and sometimes the superior of Descartes,

* Cartesii Geometria, p. 40.

was far more fortunate with regard to this problem, and his method of drawing tangents to curves is the same in effect that has been followed by all the geometers since his time,—while that of Descartes, which could only be valued when the other was unknown, has been long since entirely abandoned. The remainder of the second book is occupied with the consideration of the curves, which have been called the ovals of Descartes, and with some investigations concerning the centres of lenses; the whole indicating the hand of a great master, and deserving the most diligent study of those who would become acquainted with this great enlargement of mathematical science.

The third book of the geometry treats of the construction of equations by geometric curves, and it also contains a new method of resolving biquadratic equations.

The leading principles of algebra were now unfolded, and the notation was brought, from a mere contrivance for abridging common language, to a system of symbolical writing, admirably fitted to assist the mind in the exercise of thought.

The happy idea, indeed, of expressing quantity and the operations on quantity, by conventional symbols, instead of representing the first by real magnitudes, and enunciating the second in words, could not but make a great change on the nature of mathematical investigation. The language of mathematics, whatever may be its form, must always consist of two parts; the one denoting quantities simply, and the other denoting the manner in which the quantities are combined, or the operations understood to be performed on them. Geometry expresses the first of these by real magnitudes, or by what may be called natural signs; a line by a line, an angle by an angle, an area by an area, &c.; and it describes the latter by words. Algebra, on the other hand, denotes both quantity, and the operations on quantity, by the same system of conventional symbols. Thus, in the expression $x^3 - a x^2 + b^3 = o$, the letters a, b, x, denote quantities, but the terms x^3, ax^2, &c. denote certain operations performed on those quantities, as well as the quantities themselves; x^3 is the quantity x raised to the cube; and ax^2 the same quantity x raised to the square, and then multiplied into a, &c.; the combination, by addition or subtraction, being also expressed by the signs $+$ and $-$.

Now, it is when applied to this latter purpose that the algebraic language possesses such exclusive excellence. The mere magnitudes themselves might be represented by figures, as in geometry, as well as in any way whatever; but the operations they are to be subjected to, if described in words, must be set before the mind slowly, and in succession, so that the impression is weakened, and the clear apprehension rendered difficult. In the algebraic expression, on the other hand, so much meaning is concentrated into a narrow space, and the impression made by all the parts is so simultaneous, that nothing can be more favourable to the exertion of the reasoning powers, to the continuance of their action, and their security against error. Another advantage resulting from the use of the same notation, consists in the reduction of all the different relations among quantities to the simplest of those relations, that of equality, and the expression of it by equations. This gives a great facility of generalization. and of comparing quantities with one another. A third arises from the substitution of the arithmetical operations of multiplication and division, for the geometrical method of the composition and resolution of ratios. Of the first of these, the idea is so clear, and the work so simple; of the second, the idea is comparatively so obscure, and the process so complex, that the substitution of the former for the latter could not but be accompanied with great advantage. This is, indeed, what constitutes the great difference in practice between the algebraic and the geometric method of treating quantity. When the quantities are of a complex nature, so as to go beyond what in algebra is called the third power, the geometrical expression is so circuitous and involved, that it renders the reasoning most laborious and intricate. The great facility of generalization in algebra, of deducing one thing from another, and of adapting the analysis to every kind of research, whether the quantities be constant or variable, finite or infinite, depends on this principle more than any other. Few of the early algebraists seem to have been aware of these advantages.

The use of the signs *plus* and *minus* has given rise to some dispute. These signs were at first used the one to denote addition, the other subtraction, and for a long time were applied to no other purpose. But as, in the multiplication of a quantity, consisting of parts connected by those signs, into another quantity similarly composed, it was always found, and could be universally demonstrated, that, in uniting the particular products of which the total was made up, those of which both the factors had the sign *minus* before them, must be added into one sum with those of which all the factors had the sign *plus*; while those of which one of the factors had the sign *plus*, and the other the sign *minus*, must be subtracted from the same, —this general rule came to be more simply expressed by saying, that in multiplication like signs gave *plus*, and that unlike signs gave *minus*.

Hence the signs *plus* and *minus* were considered, not as merely denoting the relation of one quantity to another placed before it, but, by a kind of *fiction*, they were considered as denoting qualities inherent in the quantities to the names of which they were prefixed. This fiction was found extremely useful, and it was evident that no error could arise from it. It was necessary to have a rule for determining the sign belonging to a product, from the signs of the factors composing that product, independently of every other consideration; and this was precisely the purpose for which the above fiction was introduced. So necessary is this rule in the generalizations of algebra, that we meet with it in Diophantus, notwithstanding the imperfection of the language he employed; for he states, that Λειψις into Λειψις gives Ὕπαρξις, &c. The reduction, therefore, of the operations on quantity to an arithmetical form, necessarily involves this use of the signs *plus* or *minus*; that is, their application to denote something like absolute qualities in the objects they collect together. The attempts to free algebra from this use of the signs have of course failed, aud must ever do so, if we would preserve to that science the extent and facility of its operations.

Even the most scrupulous purist in mathematical language must admit, that no real error is ever introduced by employing the signs in this most abstract sense. If the equation $x^3+px^2+qx+r=o$, be said to have one positive and two negative roots, this is certainly as exceptionable an application of the term *negative*, as any that can be proposed; yet, in reality, it means nothing but this intelligible and simple truth, that $x^3+px^2+qx+r=(x-a)(x+b)(x+c)$; or that the former of these quantities is produced by the multiplication of the three binomial factors, $x-a$, $x+b$, $x+c$. We might say the same nearly as to imaginary roots; they show that the simple factors cannot be found, but that the quadratic factors may be found; and they also point out the means of discovering them.

The aptitude of these same signs to denote contrariety of position among geometric magnitudes, makes the foregoing application of them infinitely more extensive and more indispensable.

From the same source arises the great simplicity introduced into many of the theorems and rules of the mathematical sciences. Thus, the rule for finding the latitude of a place from the sun's meridian altitude, if we employ the signs *plus* and *minus* for indicating the position of the sun and of the place relatively to the equator, is enunciated in one simple proposition, which includes every case, without any thing either complex or ambiguous. But if this is not done,—if the signs *plus* and *minus* are not employed, there must be at least two rules, one when the sun and place are on the same side of the equator, and another when they are on different sides. In the more complicated calculations of spherical trigonometry, this holds still more remarkably. When one would accommodate such rules to those who are unacquainted with the use of the algebraic signs, they are perhaps not to be expressed in less than four, or even six different propositions; whereas, if the use of these signs is supposed, the whole is comprehended in a single sentence. In such cases, it is obvious that both the memory and understanding derive great advantage from the use of the signs, and profit by a simplification, which is the work entirely of the algebraic language, and cannot be imitated by any other.

That I might not interrupt the view of improvements so closely connected with one another, I have passed over one of the discoveries, which does the greatest honour to the seventeenth century, and which took place near the beginning of it.

As the accuracy of astronomical observation had been continually advancing, it was necessary that the correctness of trigonometrical calculation, and of course its difficulty, should advance in the same proportion. The sines and tangents of angles could not be expressed with sufficient correctness without decimal fractions, extending to five or six places below unity, and when to three such numbers a fourth proportional was to be found, the work of multiplication and division became extremely laborious. Accordingly, in the end of the sixteenth century, the time and labour consumed in such calculations had become excessive, and were felt as extremely burdensome by the mathematicians and astronomers all over Europe. Napier of

Merchiston, whose mind seems to have been peculiarly turned to arithmetical researches, and who was also devoted to the study of astronomy, had early sought for the means of relieving himself and others from this difficulty. He had viewed the subject in a variety of lights, and a number of ingenious devices had occurred to him, by which the tediousness of arithmetical operations might, more or less completely, be avoided. In the course of these attempts, he did not fail to observe, that whenever the numbers to be multiplied or divided were terms of a geometrical progression, the product or the quotient must also be a term of that progression, and must occupy a place in it pointed out by the places of the given numbers, so that it might be found from mere inspection, if the progression were far enough continued. If, for instance, the third term of the progression were to be multiplied by the seventh, the product must be the tenth, and if the twelfth were to be divided by the fourth, the quotient must be the eighth; so that the multiplication and division of such terms was reduced to the addition and subtraction of the numbers which indicated their places in the progression.

This observation, or one very similar to it, was made by Archimedes, and was employed by that great geometer to convey an idea of a number too vast to be correctly expressed by the arithmetical notation of the Greeks. Thus far, however, there was no difficulty, and the discovery might certainly have been made by men much inferior either to Napier or Archimedes. What remained to be done, what Archimedes did not attempt, and what Napier completely performed, involved two great difficulties. It is plain, that the resource of the geometrical progression was sufficient, when the given numbers were terms of that progression; but if they were not, it did not seem that any advantage could be derived from it. Napier, however, perceived, and it was by no means obvious, that all numbers whatsoever might be inserted in the progression, and have their places assigned in it. After conceiving the possibility of this, the next difficulty was, to discover the principle, and to execute the arithmetical process, by which these places were to be ascertained. It is in these two points that the peculiar merit of his invention consists; and at a period when the nature of series, and when every other resource of which he could avail himself were so little known, his success argues a depth and originality of thought which, I am persuaded, have rarely been surpassed.

The way in which he satisfied himself that all numbers might be intercalated between the terms of the given progression, and by which he found the places they must occupy, was founded on a most ingenious supposition,—that of two points describing two different lines, the one with a constant velocity, and the other with a velocity always increasing in the ratio of the space the point had already gone over: the first of these would generate magnitudes in arithmetical, and the second magnitudes in geometrical progression. It is plain, that all numbers whatsoever would find their places among the magnitudes so generated; and, indeed, this view of the subject is as simple and profound as any which, after two hundred years, has yet presented itself to mathematicians. The mode of deducing the results has been simplified; but it can hardly be said that the principle has been more clearly developed.

I need not observe, that the numbers which indicate the places of the terms of the geometrical progression are called by Napier the *logarithms* of those terms.

Various systems of logarithms, it is evident, may be constructed according to the geometrical progression assumed; and of these, that which was first contrived by Napier, though the simplest, and the foundation of the rest, was not so convenient for the purposes of calculation, as one which soon afterwards occurred, both to himself and his friend Briggs, by whom the actual calculation was performed. The new system of logarithms was an improvement, practically considered; but in as far as it was connected with the principle of the invention, it is only of secondary consideration. The original tables had been also somewhat embarrassed by too close a connection between them and trigonometry. The new tables were free from this inconvenience.

It is probable, however, that the greatest inventor in science was never able to do more than to accelerate the progress of discovery, and to anticipate what time, "the author of authors," would have gradually brought to light. Though logarithms had not been invented by Napier, they would have been discovered in the progress of the algebraic analysis, when the arithmetic of powers and exponents, both integral and fractional, came to be fully understood. The idea of considering all numbers, as powers of one given number, would then have readily occurred, and the doctrine of series would have greatly facilitated the calculations which it was necessary to undertake. Napier had none of these advantages, and they were all supplied by the resources of his own mind. Indeed, as there never was any invention for which the state of knowledge had less prepared the way, there never was any where more merit fell to the share of the inventor.

His good fortune, also, not less than his great sagacity, may be remarked. Had the invention of logarithms been delayed to the end of the seventeenth century, it would have come about without effort, and would not have conferred on the author the high celebrity which Napier so justly derives from it. In another respect he has also been fortunate. Many inventions have been eclipsed or obscured by new discoveries; or they have been so altered by subsequent improvements, that their original form can hardly be recognised, and, in some instances, has been entirely forgotten. This has almost always happened to the discoveries made at an early period in the progress of science, and before their principles were fully unfolded. It has been quite otherwise with the invention of logarithms, which came out of the hands of the author so perfect, that it has never received but one material improvement, that which it derived, as has just been said, from the ingenuity of his friend in conjunction with his own. Subsequent improvements in science, instead of offering any thing that could supplant this invention, have only enlarged the circle to which its utility extended. Logarithms have been applied to numberless purposes, which were not thought of at the time of their first construction. Even the sagacity of their author did not see the immense fertility of the principle he had discovered; he calculated his tables merely to facilitate arithmetical, and chiefly trigonometrical computation, and little imagined that he was at the same time constructing a scale whereon to measure the density of the strata of the atmosphere, and the heights of mountains; that he was actually computing the areas and the lengths of innumerable curves, and was preparing for a calculus which was yet to be discovered, many of the most refined and most valuable of its resources. Of Napier, therefore, if of any man, it may safely be pronounced, that his name will never be eclipsed by any one more conspicuous, or his invention superseded by any thing more valuable.

As a geometrician, Napier has left behind him a noble monument in the two trigonometrical theorems, which are known by his name, and which appear first to have been communicated in writing to Cavalieri, who has mentioned them with great eulogy.* They are theorems not a little difficult, and of much use, as being particularly adapted to logarithmic calculation. They were published in the Canon Mirificus Logarithmorum, at Edinburgh, in 1614.†

---

* Wallis, Opera Math. Tom. II. p. 875.

† A reprint of the Canon Mirificus, from the original edition, is given in the 6th Volume of the great Thesaurus, in which Baron Maseres, with his usual zeal and intelligence, has collected and illustrated everything of importance that has been written on the subject of logarithms. See Scriptores Logarithmici, 4to. Vol. VI. p. 475.

SECTION II.

EXPERIMENTAL INVESTIGATION.

IN this section I shall begin with a short view of the state of Physical Knowledge before the introduction of the Inductive Method; I shall next endeavour to explain that method by an analysis of the *Novum Organum;* and shall then inquire how far the principles established in that work have actually contributed to the advancement of Natural Philosophy.

1. ANCIENT PHYSICS.

Though the phenomena of the material world could not but early excite the curiosity of a being who, like man, receives his strongest impressions from without, yet an accurate knowledge of those phenomena, and their laws, was not to be speedily acquired. The mere extent and variety of the objects were, indeed, such obstacles to that acquisition, as could not be surmounted but in the course of many ages. Man could not at first perceive from what point he must begin his inquiries, in what direction he must carry them on, or by what rules he must be guided. He was like a traveller going forth to explore a vast and unknown wilderness, in which a multitude of great and interesting objects presented themselves on every side, while there was no path for him to follow, no rule to direct his survey, and where the art of observing, and the instruments of observation, must equally be the work of his own invention. In these circumstances, the selection of the objects to be studied was the effect of instinct rather than of reason, or of the passions and emotions more than of the understanding. When things new and unlike those which occurred in the course of every day's experience presented themselves, they excited wonder or surprise, and created an anxiety to discover some principle which might connect them with the appearances commonly observed. About these last, men felt no desire to be farther informed; but when the common order of things was violated, and something new or singular was produced, they began to examine into the fact, and attempted to inquire into the cause. Nobody sought to know why a stone fell to the ground, why smoke ascended, or why the stars revolved round the earth. But if a fiery meteor shot across the heavens,—if the flames of a volcano burst forth,—or if an earthquake shook the foundations of the world, terror and curiosity were both awakened; and when the former emotion had subsided, the latter was sure to become active. Thus, to trace a resemblance between the events with which the observer was most familiar, and those to which he was less accustomed, and which had excited his wonder, was the first object of inquiry, and produced the first advances towards generalization and philosophy. *

This principle, which it were easy to trace, from tribes the most rude and barbarous, to nations the most highly refined, was what yielded the first attempts toward classification and arrangement, and enabled man, out of individuals, subject to perpetual change, to form certain fixed and permanent objects of knowledge,—the species, genera, orders, and classes, into which he has distributed these individuals. By this effort of mental abstraction, he has created to himself a new and intellectual world, free from those changes and vicissitudes to which all material things are destined. This, too, is a work not peculiar to the philosopher, but, in a certain degree, is performed by every man who compares one thing with another, and who employs the terms of ordinary language.

---

* La maraviglia
Dell' ignoranza é la figlia,
E del sapere
La madre.

Another great branch of knowledge is occupied, not about the mere arrangement and classification of objects, but about events or changes, the laws which those changes observe, and the causes by which they are produced. In a science, which treated of events and of change, the nature and properties of motion came of course to be studied, and the ancient philosophers naturally enough began their inquiries with the definition of motion, or the determination of that in which it consists. Aristotle's definition is highly characteristical of the vagueness and obscurity of his physical speculations. He calls motion "the act of a being in power, as far as in power,"—words to which it is impossible that any distinct idea can ever have been annexed.

The truth is, however, that the best definition of motion can be of very little service in physics. Epicurus defined it to be the "change of place," which is, no doubt, the simplest and best definition that can be given; but it must, at the same time, be confessed, that neither he, nor the moderns who have retained his definition, have derived the least advantage from it in their subsequent researches. The properties, or, as they are called, the laws of motion, cannot be derived from mere definition; they must be sought for in experience and observation, and are not to be found without a diligent comparison, and scrupulous examination of facts.

Of such an examination, neither Aristotle, nor any other of the ancients, ever conceived the necessity, and hence those laws remained quite unknown throughout all antiquity.

When the laws of motion were unknown, the other parts of natural philosophy could make no great advances. Instead of conceiving that there resides in *body* a natural and universal tendency to persevere in the same state, whether of rest, or of motion, they believed that terrestrial bodies tended *naturally* either to fall to the ground, or to ascend from it, till they attained their own place; but that, if they were impelled by an oblique force, then their motion became *unnatural* or *violent*, and tended continually to decay. With the heavenly bodies, again, the natural motion was circular and uniform, eternal in its course, but perpetually varying in its direction. Thus, by the distinction between natural and violent motion among the bodies of the earth, and the distinction between what we may call the laws of motion in terrestrial and celestial bodies, the ancients threw into all their reasonings upon this fundamental subject a confusion and perplexity, from which their philosophy never was delivered.

There was, however, one part of physical knowledge in which their endeavours were attended with much better success, and in which they made important discoveries. This was in the branch of

Mechanics, which treats of the action of forces *in equilibrio*, and producing not motion but rest;—a subject which may be understood, though the laws of motion are unknown.

The first writer on this subject is Archimedes. He treated of the lever, and of the centre of gravity, and has shown that there will be an equilibrium between two heavy bodies connected by an inflexible rod or lever, when the point in which the lever is supported is so placed between the bodies, that their distances from it are inversely as their weights. Great ingenuity is displayed in this demonstration; and it is remarkable, that the author borrows no principle from experiment, but establishes his conclusion entirely by reasoning *a priori*. He assumes, indeed, that equal bodies, at the ends of the equal arms of a lever, will balance one another; and also, that a cylinder, or parallelepiped of homogeneous matter, will be balanced about its centre of magnitude. These, however, are not inferences from experience; they are, properly speaking, conclusions deduced from the principle of the *sufficient reason*.

The same great geometer gave a beginning to the science of Hydrostatics, and discovered the law which determines the loss of weight sustained by a body on being immersed in water, or in any other fluid. His demonstration rests on a principle, which he lays down as a postulatum, that, in water, the parts which are less pressed are always ready to yield in any direction to those that are more pressed, and from this, by the application of mathematical reasoning, the whole theory of floating bodies is derived. The above is the same principle on which the modern writers on hydrostatics proceed; they give it not as a postulatum, but as constituting the definition of a fluid.

Archimedes, therefore, is the person who first made the application of mathematics to natural philosophy. No individual, perhaps, ever laid the foundation of more great discoveries than that geometer, of whom Wallis has said with so much truth, " Vir stupendæ sagacitatis, qui prima fundamenta posuit inventionum feré omnium in quibus promovendis ætas nostra gloriatur."

The mechanical inquiries, begun by the geometer of Syracuse, were extended by Ctesibius and Hero; by Anthemius of Tralles; and, lastly, by Pappus Alexandrinus. Ctesibius and Hero were the first who analyzed mechanical engines, reducing them all to combinations of five simple mechanical contrivances, to which they gave the name of Δυναμεις, or Powers, the same which they retain at the present moment.

Even in mechanics, however, the success of these investigations was limited; and failed in those cases where the resolution of forces is necessary, that principle being then entirely unknown. Hence the force necessary to sustain a body on an inclined plane, is incorrectly determined by Pappus, and serves to mark a point to which the mechanical theories of antiquity did not extend.

In another department of physical knowledge, Astronomy, the endeavours of the ancients were also accompanied with success. I do not here speak of their astronomical theories, which were, indeed, very defective, but of their discovery of the apparent motions of the heavenly bodies, from the observations begun by Hipparchus, and continued by Ptolemy. In this their success was great; and while the earth was supposed to be at rest, and while the instruments of observation had but a very limited degree of accuracy, a nearer approach to the truth was probably not within the power of human ingenuity. Mathematical reasoning was very skilfully applied, and no men whatever, in the same circumstances, are likely to have performed more than the ancient astronomers. They succeeded, because they were observers, and examined carefully the motions which they treated of. The philosophers, again, who studied the motion of terrestrial bodies, either did not observe at all, or observed so slightly, that they could obtain no accurate knowledge, and in general they knew just enough of the facts to be misled by them.

The opposite ways which the ancients thus took to study the Heavens and the Earth, observing the one, and dreaming, as one may say, over the other, though a striking inconsistency, is not difficult to be explained.

No information at all could be obtained in astronomy, without regular and assiduous observation, and without instruments capable of measuring angles, and of measuring time, either directly or indirectly. The steadiness and regularity of the celestial motions seemed to invite the most scrupulous attention. On the other hand, as terrestrial objects were always at hand, and spontaneously falling under men's view, it seemed unnecessary to take much trouble to become acquainted with them, and as for employing measures, their irregularity appeared to render every idea of such a proceeding nugatory. The Aristotelian philosophy particularly favoured this prejudice, by representing the earth, and all things on its surface, as full of irregularity and confusion, while the principles of heat and cold, dryness and moisture, were in a state of perpetual warfare. The unfortunate division of motion into natural and violent, and the distinction, still more unfortunate, between the properties of motion and of body, in the heavens and on the earth, prevented all intercourse between the astronomer and the naturalist, and all transference of the maxims of the one to the speculations of the other.

Though, on account of this inattention to experiment, nothing like the true system of natural philosophy was known to the ancients, there are, nevertheless, to be found in their writings many brilliant conceptions, several fortunate conjectures, and gleams of the light which was afterwards to be so generally diffused.

Anaxagoras and Empedocles, for example, taught that the moon shines by light borrowed from the sun, and were led to that opinion, not only from the phases of the moon, but from its light being weak, and unaccompanied by heat. That it was a habitable body, like the earth, appears to be a doctrine as old as Orpheus; some lines, ascribed to that poet, representing the moon as an earth, with mountains and cities on its snrface.

Democritus supposed the spots on the face of the moon to arise from the inequalities of the surface, and from the shadows of the more elevated parts projected on the plains. Every one knows how conformable this is to the discoveries made by the telescope.

Plutarch considers the velocity of the moon's motion as the cause which prevents that body from falling to the earth, just as the motion of a stone in a sling prevents it from falling to the ground. The comparison is, in a certain degree, just, and clearly implies the notion of centrifugal force; and gravity may also be considered as pointed at for the cause which gives the moon a tendency to the earth. Here, therefore, a foundation was laid for the true philosophy of the celestial motions; but it was laid without effect. It was merely the conjecture of an ingenious mind, wandering through the regions of possibility, guided by no evidence, and having no principle which could give stability to its opinions. Democritus, and the authors of that physical system which Lucretius has so beautifully illustrated, were still more fortunate in some of their conjectures. They taught that the Milky Way is the light of a great number of small stars, very close to one another; a magnificent conception, which the latest improvements of the telescope have fully verified. Yet, as if to convince us that they derived this knowledge from no pure or certain source, the same philosophers maintained, that the sun and the moon are bodies no larger than they appear to us to be.

Very just notions concerning comets were entertained by some of the ancients. The Chaldeans considered those bodies as belonging to the same order with the planets; and this was also the opinion of Anaxagoras, Pythagoras, and Democritus. The remark of Seneca on this subject is truly philosophical, and contains a prediction which has been fully accomplished: "Why do we wonder that comets, which are so rare a spectacle in the world, observe laws which to us are yet unknown, and that the beginning and end of motions, so seldom observed, are not yet fully understood?"—*Veniet tempus, quo ista quæ nunc latent, in lucem dies extrahat, et longioris ævi diligentia : ad inquisitionem tantorum ætas una non sufficit. Veniet tempus, quo posteri nostri tam aperta nos nescisse mirentur.* *

It was, however, often the fate of such truths to give way to error. The comets, which these ancient philosophers had ranked so justly with the stars, were degraded by Aristotle into meteors floating in the earth's atmosphere; and this was the opinion concerning them which ultimately prevailed.

But, notwithstanding the above, and a few other splendid conceptions which shine through the obscurity of the ancient physics, the system, taken on the whole, was full of error and inconsistency. Truth and falsehood met almost on terms of equality; the former separated from its root, experience, found no preference above the latter; to the latter, in fact, it was generally forced to give way, and the dominion of error was finally established.

One ought to listen, therefore, with caution to the encomiums sometimes bestowed on the philosophy of those early ages. If these encomiums respected only the talents, the genius, the taste of the great masters of antiquity, we would subscribe to them without any

* Nat. Quæst. Lib. vii. c. 25.

apprehension of going beyond the truth. But if they extend to the methods of philosophizing, and the discoveries actually made, we must be excused for entering our dissent, and exchanging the language of panegyric for that of apology. The infancy of science could not be the time when its attainments were the highest; and, before we suffer ourselves to be guided by the veneration of antiquity, we ought to consider in what real antiquity consists. With regard to the progress of knowledge and improvement, "*we* are more ancient than those who went before us." * The human race has now more experience than in the generations that are past, and of course may be expected to have made higher attainments in science and philosophy. Compared with natural philosophy, as it now exists, the ancient physics are rude and imperfect. The speculations contained in them are vague and unsatisfactory, and of little value, but as they elucidate the history of the errors and illusions to which the human mind is subject. Science was not merely stationary, but often retrograde; the earliest opinions were frequently the best; and the reasonings of Democritus and Anaxagoras were in many instances more solid than those of Plato and Aristotle. Extreme credulity disgraced the speculations of men who, however ingenious,

---

* Bacon.

were little acquainted with the laws of nature, and unprovided with the great criterion by which the evidence of testimony can alone be examined. Though observations were sometimes made, experiments were never instituted; and philosophers, who were little attentive to the facts which spontaneously offered, did not seek to increase their number by artificial combinations. Experience, in those ages, was a light which darted a few tremulous and uncertain rays on some small portions of the field of science, but men had not acquired the power over that light which now enables them to concentrate its beams, and to fix them steadily on whatever object they wish to examine. This power is what distinguishes the modern physics, and is the cause why later philosophers, without being more ingenious than their predecessors, have been infinitely more successful in the study of nature.

2. Novum Organum.

The defects which have been ascribed to the ancient physics were not likely to be corrected in the course of the middle ages. It is true, that, during those ages, a science of pure experiment had made its appearance in the world, and might have been expected to remedy the greatest of these defects, by turning the attention of philosophers to experience and observation. This effect, however, was far from being immediately produced; and none who professed to be in search of truth ever wandered over the regions of fancy, in paths more devious and eccentric, than the first experimenters in chemistry. They had become acquainted with a series of facts so unlike to any thing already known, that the ordinary principles of belief were shaken or subverted, and the mind laid open to a degree of credulity far beyond any with which the philosophers of antiquity could be reproached. An unlooked-for extension of human power had taken place; its limits were yet unknown; and the boundary between the possible and the impossible was no longer to be distinguished. The adventurers in an unexplored country, given up to the guidance of imagination, pursued objects which the kindness, no less than the wisdom of nature, have rendered unattainable by man; and in their speculations peopled the air, the earth, and all the elements, with spirits and genii, the invisible agents destined to connect together all the facts which they knew, and all those which they hoped to discover. Chemistry, in this state, might be said to have an elective attraction for all that was most absurd and extravagant in the other parts of knowledge; alchemy was its immediate offspring, and it allied itself in succession with the dreams of the

Cabbalists, the Rosicrucians, and the Theosophers. Thus a science, founded in experiment, and destined one day to afford such noble examples of its use, exhibited for several ages little else than a series of illusory pursuits, or visionary speculations, while now and then a fact was accidentally discovered.

Under the influence of these circumstances arose Paracelsus, Van Helmont, Fludde, Cardan, and several others, conspicuous no less for the weakness than the force of their understandings: men who united extreme credulity, the most extravagant pretensions, and the most excessive vanity, with considerable powers of invention, a complete contempt for authority, and a desire to consult experience; but destitute of the judgment, patience, and comprehensive views, without which the responses of that oracle are never to be understood. Though they appealed to experience, and disclaimed subjection to the old legislators of science, they were in too great haste to become legislators themselves, and to deduce an explanation of the whole phenomena of nature from a few facts, observed without accuracy, arranged without skill, and never compared or confronted with one another. Fortunately, however, from the turn which their inquiries had taken, the ill done by them has passed away, and the good has become permanent. The reveries of Paracelsus have disappeared, but his application of chemistry to pharmacy has conferred a lasting benefit on the world. The *Archæus* of Van Helmont, and the army of spiritual agents with which the discovery of elastic fluids had filled the imagination of that celebrated empiric, are laughed at, or forgotten; but the fluids which he had the sagacity to distinguish, form, at the present moment, the connecting principles of the new chemistry.

Earlier than any of the authors just named, but in a great measure under the influence of the same delusions, Roger Bacon appears to have been more fully aware than any of them of the use of experiment, and of mathematical reasoning, in physical and mechanical inquiries. But, in the thirteenth century, an appeal from the authority of the schools, even to nature herself, could not be made with impunity. Bacon, accordingly, incurred the displeasure both of the University and of the Church, and this forms one of his claims to the respect of posterity, as it is but fair to consider persecution inflicted by the ignorant and bigoted as equivalent to praise bestowed by the liberal and enlightened.

Much more recently, Gilbert, in his treatise on the Magnet, had given an example of an experimental inquiry, carried on with more correctness, and more enlarged views, than had been done by any of his predecessors. Nevertheless, in the end of the sixteenth century, it might still be affirmed, that the situation of the great avenue to knowledge was fully understood by none, and that its existence, to the bulk of philosophers, was utterly unknown.

It was about this time that Francis Bacon (Lord Verulam) began to turn his powerful and creative mind to contemplate the state of human knowledge, to mark its imperfections, and to plan its improvement. One of the considerations which appears to have impressed his mind most forcibly, was the vagueness and uncertainty of all the physical speculations then existing, and the entire want of connection between the sciences and the arts.

Though these two things are in their nature so closely united, that the same truth which is a principle in science, becomes a rule in art, yet there was at that time hardly any practical improvement which had arisen from a theoretical discovery. The natural alliance between the knowledge and the power of man seemed entirely interrupted; nothing was to be seen of the mutual support which they ought to afford to one another; the improvement of art was left to the slow and precarious operation of chance, and that of science to the collision of opposite opinions.

" But whence," said Bacon, " can arise such vagueness and sterility in all the physical systems which have hitherto existed in the world? It is not certainly from any thing in nature itself; for the steadiness and regularity of the laws by which it is governed clearly mark them out as objects of certain and precise knowledge. Neither can it arise from any want of ability in those who have pursued such inquiries, many of whom have been men of the highest talent and genius of the ages in which they lived; and it can, therefore, arise from nothing else but the perverseness and insufficiency of the methods that have been pursued. Men have sought to make a world from their own conceptions, and to draw from their own minds all the materials which they employed; but if, instead of doing so, they had consulted experience and observation, they would have had facts, and not opinions, to reason about, and might have ultimately arrived at the knowledge of the laws which govern the material world."

"As things are at present conducted," he adds, "a sudden transition is made from sensible objects and particular facts to general propositions, which are accounted principles, and round which, as round so many fixed poles, disputation and argument continually revolve. From the propositions thus hastily assumed, all things are derived, by a process compendious and precipitate, ill suited to discovery, but wonderfully accommodated to debate. The way that promises success is the reverse of this. It requires that we should generalize slowly, going from particular things to those that are but one step more general; from those to others of still greater extent, and so on to such as are universal. By such means, we may hope to arrive at principles, not vague and obscure, but luminous and well-defined, such as nature herself will not refuse to acknowledge."

Before laying down the rules to be observed in this inductive process, Bacon proceeds to enumerate the causes of error,—the *Idols*, as he terms them, in his figurative language, or false divinities to which the mind had so long been accustomed to bow. He considered this enumeration as the more necessary, that the same idols were likely to return, even after the reformation of science, and to avail themselves of the real discoveries that might have been made, for giving a colour to their deceptions.

These idols he divides into four classes, to which he gives names, fantastical, no doubt, but, at the same time, abundantly significant.

Idola Tribus,	Idols of the Tribe.
—— Specus,	—— of the Den.
—— Fori,	—— of the Forum.
—— Theatri,	—— of the Theatre.

1. The *idols of the tribe*, or of the race, are the causes of error founded on human nature in general, or on principles common to all mankind. "The mind," he observes, "is not like a plane

mirror, which reflects the images of things exactly as they are; it is like a mirror of an uneven surface, which combines its own figure with the figures of the objects it represents."*

Among the idols of this class, we may reckon the propensity which there is in all men to find in nature a greater degree of order, simplicity, and regularity, than is actually indicated by observation. Thus, as soon as men perceived the orbits of the planets to return into themselves, they immediately supposed them to be perfect circles, and the motion in those circles to be uniform; and to these hypotheses, so rashly and gratuitously assumed, the astronomers and mathematicians of all antiquity laboured incessantly to reconcile their observations.

The propensity which Bacon has here characterized so well, is the same that has been, since his time, known by the name of the *spirit of system.* The prediction, that the sources of error would return, and were likely to infest science in its most flourishing condition, has been fully verified with respect to this illusion, and in the case of sciences which had no existence at the time when Bacon wrote. When it was ascertained, by observation, that a considerable part of the earth's surface consists of minerals, disposed in horizontal strata, it

* Novum Organum, Lib. i. Aph. 41.

was immediately concluded, that the whole exterior crust of the earth is composed, or has been composed, of such strata, continued all round without interruption; and on this, as on a certain and general fact, entire theories of the earth have been constructed.

There is no greater enemy which science has to struggle with than this propensity of the mind; and it is a struggle from which science is never likely to be entirely relieved; because, unfortunately, the illusion is founded on the same principle from which our love of knowledge takes its rise.

2. The *idols of the den* are those that spring from the peculiar character of the individual. Besides the causes of error which are common to all mankind, each individual, according to Bacon, has his own dark cavern or den, into which the light is imperfectly admitted, and in the obscurity of which a tutelary idol lurks, at whose shrine the truth is often sacrificed.

One great and radical distinction in the capacities of men is derived from this, that some minds are best adapted to mark the differences, others to catch the resemblances, of things. Steady and profound understandings are disposed to attend carefully, to proceed slowly, and to examine the most minute differences; while those that are sublime and active are ready to lay hold of the slightest resemblances. Each of these easily runs into excess; the one by catching continually at distinctions, the other at affinities.

The studies, also, to which a man is addicted, have a great effect in influencing his opinions. Bacon complains, that the chemists of his time, from a few experiments with the furnace and the crucible, thought that they were furnished with principles sufficient to explain the structure of the universe; and he censures Aristotle for having depraved his physics so much with his dialectics, as to render the former entirely a science of words and controversy. In like manner, he blames a philosopher of his own age, Gilbert, who had studied magnetism to good purpose, for having proceeded to form out of it a general system of philosophy. Such things have occurred in every period of science. Thus electricity has been applied to explain the motion of the heavenly bodies; and, of late, galvanism and electricity together have been held out as explaining, not only the affinities of chemistry, but the phenomena of gravitation, and the laws of vegetable and animal life. It were a good caution for a man who studies nature, to distrust those things with which he is particularly conversant, and which he is accustomed to contemplate with pleasure.

3. The *idols of the forum* are those that arise out of the commerce or intercourse of society, and especially from language, or the means by which men communicate their thoughts to one another.

Men believe that their thoughts govern their words; but it also happens, by a certain kind of reaction, that their words frequently govern their thoughts. This is the more pernicious, that words, being generally the work of the multitude, divide things according to the lines most conspicuous to vulgar apprehensions. Hence, when words are examined, few instances are found in which, if at all abstract, they convey ideas tolerably precise and well defined. For such imperfections there seems to be no remedy, but by having recourse to particular instances, and diligently comparing the meanings of words with the external archetypes from which they are derived.

4. The *idols of the theatre* are the last, and are the deceptions which have taken their rise from the systems or dogmas of the different schools of philosophy. In the opinion of Bacon, as many of these systems as had been invented, so many representations of imaginary worlds had been brought upon the stage. Hence the name of *idola theatri.* They do not enter the mind imperceptibly like the other three; a man must labour to acquire them, and they are often the result of great learning and study.

"Philosophy," said he, "as hitherto pursued, has taken much from a few things, or a little from a great many; and, in both cases, has too narrow a basis to be of much duration or utility." The

Aristotelian philosophy is of the latter kind; it has taken its principles from common experience, but without due attention to the evidence or the precise nature of the facts; the philosopher is left to work out the rest from his own invention. Of this kind, called by Bacon the *sophistical*, were almost all the physical systems of antiquity.

When philosophy takes all its principles from a few facts, he calls it *empirical*,—such as was that of Gilbert, and of the chemists.

It should be observed, that Bacon does not charge the physics of antiquity with being absolutely regardless of experiment. No system, indeed, however fantastical, has ever existed, to which that reproach could be applied in its full extent; because, without some regard to fact, no theory can ever become in the least degree plausible. The fault lies not, therefore, in the absolute rejection of experience, but in the unskilful use of it; in taking up principles lightly from an inaccurate and careless observation of many things; or, if the observations have been more accurate, from those made on a few facts, unwarrantably generalized.

Bacon proceeds to point out the circumstances, in the history of the world, which had hitherto favoured these perverse modes of philosophizing. He observes, that the periods during which science had been cultivated were not many, nor of long duration. They might be reduced to three; the first with the Greeks; the second with the Romans; and the third with the western nations, after the revival of letters. In none of all these periods had much attention been paid to natural philosophy, the great parent of the sciences.

With the Greeks, the time was very short during which physical science flourished in any degree. The seven Sages, with the exception of Thales, applied themselves entirely to morals and politics; and in later times, after Socrates had brought down philosophy from the heavens to the earth, the study of nature was generally abandoned. In the Roman republic, the knowledge most cultivated, as might be expected among a martial and ambitious people, was such as had a direct reference to war and politics. During the empire, the introduction and establishment of the Christian religion drew the attention of men to theological studies, and the important interests which were then at stake left but a small share of talent and ability to be occupied in inferior pursuits. The corruptions which followed, and the vast hierarchy which assumed the command both of the sword and the sceptre, while it occupied and enslaved the minds of men, looked with suspicion on sciences which could not easily be subjected to its control.

At the time, therefore, when Bacon wrote, it might truly be said, that a small portion even of the learned ages, and of the abilities of learned men, had been dedicated to the study of Natural Philosophy. This served, in his opinion, to account for the imperfect state in which he found human knowledge in general; for he thought it certain, that no part of knowledge could attain much excellence without having its foundation laid in physical science.

He goes on to observe, that the end and object of knowledge had been very generally mistaken; that many, instead of seeking through it to improve the condition of human life, by new inventions and new resources, had aimed only at popular applause, and had satisfied themselves with the knowledge of words more than of things: while others, who were exceptions to this rule, had gone still farther wrong, by directing their pursuits to objects imaginary and unattainable. The alchemists, for example, alternately the dupes of their own credulity and of their own imposture, had amazed and tormented the world with hopes which were never to be realized. Others, if possible more visionary, had promised to prolong life, to extinguish disease and infirmity, and to give man a command over the world of spirits, by means of mystic incantations. "All this," says he, "is the mere boasting of ignorance; for, when the knowledge of nature shall be rightly pursued, it will lead to discoveries that will as far excel the pretended powers of magic, as the real exploits of Cæsar and Alexander exceed the fabulous adventures of Arthur of Britain, or Amadis of Gaul."*

Again, the reverence for antiquity, and the authority of great names, have contributed much to retard the progress of science. Indeed, the notion of antiquity which men have taken up seems to be erroneous and inconsistent. It is the duration of the world, or of the human race, as reckoned from the extremity that is past, and not from the point of time which is present, that constitutes the true antiquity to which the advancement of science may be conceived to bear some proportion ; and just as we expect more wisdom and experience in an old than in a young man, we may expect more knowledge of nature from the present than from any of the ages that are past.

"It is not to be esteemed a small matter in this estimate, that, by the voyages and travels of these later times, so much more of nature has been discovered than was known at any former period. It would, indeed, be disgraceful to mankind, if, after such tracts of the material world have been laid open, which were unknown in former times,—so many seas traversed,—so many countries explored, —so many stars discovered,— that philosophy, or the intelligible world, should be circumscribed by the same boundaries as before."

---

* Nov. Org. Lib. i. Aph. 87.

Another cause has greatly obstructed the progress of philosophy, viz. that men inquire only into the causes of rare, extraordinary, and great phenomena, without troubling themselves about the explanation of such as are common, and make a part of the general course of nature. * It is, however, certain, that no judgment can be formed concerning the extraordinary and singular phenomena of nature, without comparing them with those that are ordinary and frequent.

The laws which are every day in action, are those which it is most important for us to understand ; and this is well illustrated by what has happened in the scientific world since the time when Bacon wrote. The simple falling of a stone to the ground has been found to involve principles which are the basis of all we know in mechanical philosophy. Without accurate experiments on the descent of bodies at the surface of the earth, the objections against the earth's motion could not have been answered, the inertia of body would have remained unknown, and the nature of the force which retains the planets in their orbits could never have been investigated. Nothing, therefore, can be more out of its place than the fastidiousness of those philosophers, who suppose things to be unworthy of study, because, with respect to ordinary life, they are tri-

* Nov. Org. Lib. i. Aph. 119.

vial and unimportant. It is an error of the same sort which leads men to consider experiment, and the actual application of the hands, as unworthy of them, and unbecoming of the dignity of science. "There are some," says Bacon, "who, delighting in mere contemplation, are offended with our frequent reference to experiments and operations to be performed by the hand, things which appear to them mean and mechanical; but these men do in fact oppose the attainment of the object they profess to pursue, since the exercise of contemplation, and the construction and invention of experiments, are supported on the same principles, and perfected by the same means."

After these preliminary discussions, the great restorer of philosophy proceeds, in the second book of the Novum Organum, to describe and exemplify the nature of the induction, which he deems essential to the right interpretation of nature.

The first object must be to prepare a history of the phenomena to be explained, in all their modifications and varieties. This history is to comprehend not only all such facts as spontaneously offer themselves, but all the experiments instituted for the sake of discovery, or for any of the purposes of the useful arts. It ought to be composed with great care; the facts accurately related, and distinctly arranged; their authenticity diligently examined; those that rest on doubtful evidence, though not rejected, being noted as uncertain, with the grounds of the judgment so formed. This last is very necessary; for facts often appear incredible, only because we are ill informed, and cease to appear marvellous, when our knowledge is farther extended.

All such facts, however, as appear contrary to the ordinary course of our experience, though thus noted down and preserved, must have no weight allowed them in the first steps of investigation, and are to be used only when the general principle, as it emerges from the inductive process, serves to increase their probability.

This record of facts is what Bacon calls natural history, and it is material to take notice of the comprehensive sense in which that term is understood through all his writings. According to the arrangement of the sciences, which he has explained in his treatise on the advancement of knowledge, all learning is classed relatively to the three intellectual faculties of Memory, Reason, and Imagination. Under the first of these divisions is contained all that is merely Narration or History, of whatever kind it may be. Under the second are contained the different sciences, whether they respect the Intellectual or the Material world. Under the third are comprehended Poetry and the Fine Arts. It is with the first of these classes only that we are at present concerned. The two first divisions of it are Sacred and Civil History, the meaning of which is sufficiently understood. The third division is Natural History, which comprehends the description of the facts relative to inanimate matter, and to all animals, except man. Natural history is again subdivided into three parts: 1. The history of the phenomena of nature, which are uniform; 2. Of the facts which are anomalous or extraordinary; 3. Of the processes in the different arts.

We are not to wonder at finding the processes of the arts thus enrolled among the materials of natural history. The powers which act in the processes of nature and in those of art are precisely the same, and are only directed, in the latter case, by the intention of man, toward particular objects. In art, as Bacon elsewhere observes, man does nothing more than bring things nearer to one another, or carry them farther off; the rest is performed by nature, and, on most occasions, by means of which we are quite ignorant.

Thus, when a man fires a pistol, he does nothing but make a piece of flint approach a plate of hardened steel, with a certain velocity. It is nature that does the rest;—that makes the small red hot and fluid globules of steel, which the flint had struck off, communicate their fire to the gunpowder, and, by a process but little understood, set loose the elastic fluid contained in it; so that an explosion is produced, and the ball propelled with astonishing velocity. It is obvious that, in this instance, art only gives certain powers of nature a particular direction.

To the rules which have been given from Bacon, for the composition of natural history, I may be permitted to add this other,—that theoretical language should, as much as possible, be avoided. Appearances ought to be described in terms which involve no opinion with respect to their causes. These last are the objects of separate examination, and will be best understood if the facts are given fairly, without any dependence on what should yet be considered as unknown. This rule is very essential where the facts are in a certain degree complicated; for it is then much easier to describe with a reference to theory than without it. It is only from a skilful physician that you can expect a description of a disease which is not full of opinions concerning its cause. A similar observation might be made with respect to agriculture; and with respect to no science more than geology.

The natural history of any phenomenon, or class of phenomena, being thus prepared, the next object is, by a comparison of the different facts, to find out the cause of the phenomenon, its *form*, in the language of Bacon, or its essence. The form of any quality in body is something convertible with that quality; that is, where it exists, the quality is present, and where the quality is present, the form must be so likewise. Thus, if transparency in bodies be the thing inquired after, the form of it is something that, wherever it is found, there is transparency; and, *vice versa*, wherever there is transparency, that which we have called the form is likewise present.

The form, then, differs in nothing from the cause; only we apply the word cause where it is event or change that is the effect. When the effect or result is a permanent quality, we speak of the form or essence.

Two other objects, subordinate to *forms*, but often essential to the knowledge of them, are also occasionally subjects of investigation. These are the latent process, and the latent schematism; *latens processus, et latens schematismus*. The former is the secret and invisible progress by which sensible changes are brought about, and seems, in Bacon's acceptation, to involve the principle, since called the *law of continuity*, according to which, no change, however small, can be effected but in *time*. To know the relation between the time and the change effected in it, would be to have a perfect knowledge of the latent process. In the firing of a cannon, for example, the succession of events during the short interval between the application of the match and the expulsion of the ball, constitutes a latent process of a very remarkable and complicated nature, which, however, we can now trace with some degree of accuracy. In mechanical operations, we can often follow this process still more completely. When motion is communicated from any body to another, it is distributed through all the parts of that other, by a law quite beyond the reach of sense to perceive directly, but yet subject to investigation, and determined by a principle, which, though late of being discovered, is now generally recognised. The applications of this mechanical principle are perhaps the instances in which a latent, and, indeed, a very recondite process, has been most completely analyzed.

The latent schematism is that invisible structure of bodies, on which so many of their properties depend. When we inquire into the constitution of crystals, or into the internal structure of plants, &c. we are examining into the latent schematism. We do the same when we attempt to explain elasticity, magnetism, gravitation, &c. by any peculiar structure of bodies, or any arrangement of the particles of matter. *

In order to inquire into the *form* or cause of any thing by induction, having brought together the facts, we are to begin with considering what things are thereby excluded from the number of possible forms. This exclusion is the first part of the process of induction: it confines the field of

* Nov. Org. Lib. ii. Aph. 5, 6, &c.

hypothesis, and brings the true explanation within narrower limits. Thus, if we were inquiring into the quality which is the cause of transparency in bodies; from the fact that the diamond is transparent, we immediately exclude rarity or porosity as well as fluidity from those causes, the diamond being a very solid and dense body.

Negative instances, or those where the given *form* is wanting, are also to be collected.

That glass, when pounded, is not transparent, is a negative fact, and of considerable importance when the *form* of transparency is inquired into; also, that collections of vapour, such as clouds and fogs, have not transparency, are negative facts of the same kind. The facts thus collected, both affirmative and negative, may, for the sake of reference, be reduced into tables.

Bacon exemplifies his method on the subject of Heat; and, though his collection of facts be imperfect, his method of treating them is extremely judicious, and the whole disquisition highly interesting. * He here proposes, as an experiment, to try the reflection of the heat of opaque bodies. † He mentions also the *vitrum calendare*, or thermometer, which was just then coming into use. His reflections, after finishing his enumeration of facts,

---

* Nov. Org. Lib. ii. Aph. 18, 20, &c.

† Ibid. Aph. 11.

show how sensible he was of the imperfect state of his own knowledge. *

After a great number of exclusions have left but a few principles, common to every case, one of these is to be assumed as the cause; and, by reasoning from it synthetically, we are to try if it will account for the phenomena.

So necessary did this exclusive process appear to Bacon, that he says, " It may perhaps be competent to angels, or superior intelligences, to determine the form or essence directly, by affirmations from the first consideration of the subject. But it is certainly beyond the power of man, to whom it is only given to proceed at first by negatives, and, in the last place, to end in an *affirmative*, after the exclusion of every thing else. †

The method of induction, as laid down here, is to be considered as applicable to all investigations where experience is the guide, whether in the moral or natural world. " Some may doubt whether we propose to apply our method of investigation to natural philosophy only, or to other sciences, such as logic, ethics, politics. We answer, that we mean it to be so applied. And as the common logic, which proceeds by the syllogism, belongs not only to natural philosophy, but to all the sciences,

* Nov Org. Lib. ii. Aph. 14.

† Ibid. Aph. 15.

so our logic, which proceeds by induction, embraces every thing." *

Though this process had been pursued by a person of much inferior penetration and sagacity to Bacon, he could not but have discovered that all facts, even supposing them truly and accurately recorded, are not of equal value in the discovery of truth. Some of them show the thing sought for in its highest degree, some in its lowest; some exhibit it simple and uncombined, in others it appears confused with a variety of circumstances. Some facts are easily interpreted, others are very obscure, and are understood only in consequence of the light thrown on them by the former. This led our author to consider what he calls *Prerogativæ Instantiarum*, the comparative value of facts as means of discovery, or as instruments of investigation. He enumerates twenty-seven different species, and enters at some length into the peculiar properties of each. I must content myself, in this sketch, with describing a few of the most important, subjoining, as illustrations, sometimes the examples which the author himself has given, but more frequently such as have been furnished by later discoveries in science.

I. The first place in this classification is assigned to what are called *instantiæ solitariæ*, which are

---

* Nov. Org. Lib. i. Aph. 127.

either examples of the same quality existing in two bodies, which have nothing in common but that quality, or of a quality differing in two bodies, which are in all other respects the same. In the first instance, the bodies differ in all things but one; in the second, they agree in all but one. The *hypotheses* that in either case can be entertained, concerning the cause or *form* of the said quality, are reduced to a small number; for, in the first, they can involve none of the things in which the bodies differ; and, in the second, none of those in which they agree.

Thus, of the cause or *form* of colour now inquired into, *instantiæ solitariæ* are found in crystals, prisms of glass, drops of dew, which occasionally exhibit colour, and yet have nothing in common with the stones, flowers, and metals, which possess colour permanently, except the colour itself. Hence Bacon concludes, that colour is nothing else than a modification of the rays of light, produced, in the first case, by the different degrees of incidence; and, in the second, by the texture or constitution of the surfaces of bodies. He may be considered as very fortunate in fixing on these examples, for it was by means of them that Newton afterwards found out the composition of light.

Of the second kind of *instantiæ solitariæ*, Bacon mentions the white or coloured veins which occur in limestone or marble, and yet hardly differ in substance or in structure from the ground of the stone. He concludes, very justly, from this, that colour has not much to do with the essential properties of body.

II. The *instantiæ migrantes* exhibit some *nature* or property of body, passing from one condition to another, either from less to greater, or from greater to less; arriving nearer perfection in the first case, or verging towards extinction in the second.

Suppose the thing inquired into were the cause of whiteness in bodies; an *instantia migrans* is found in glass, which, when entire, is without colour, but, when pulverized, becomes white. The same is the case with water unbroken, and water dashed into foam. In both cases, the separation into particles produces whiteness. So also the communication of fluidity to metals by the application of heat; and the destruction of that fluidity by the abstraction of heat, are examples of both kinds of the *instantia migrans*. Instances of this kind are very powerful for reducing the cause inquired after into a narrow space, and for removing all the accidental circumstances. It is necessary, however, as Bacon * very justly remarks, that we should consider not merely the case when a certain quality is lost, and another produced, but the gra-

---

* Nov. Org. Lib. ii. Aph. 23.

dual changes made in those qualities during their migration, viz. the increase of the one, and the corresponding diminution of the other. The quantity which changes proportionally to another, is connected with it either as cause and effect, or as a collateral effect of the same cause. When, again, we find two qualities which do not increase proportionally, they afford a negative instance, and assure us that the two are not connected simply as cause and effect.

The mineral kingdom is the great theatre of the *instantiæ migrantes*, where the same *nature* is seen in all gradations, from the most perfect state, till it become entirely evanescent. Such are the shells which we see so perfect in figure and structure in limestone, and gradually losing themselves in the finer marbles, till they can no longer be distinguished. The use, also, of one such fact to explain or interpret another, is no-where so well seen as in the history of the mineral kingdom.

III. In the third place are the *instantiæ ostensivæ*, which Bacon also calls *elucescentiæ* and *predominantes*. They are the facts which show some particular *nature* in its highest state of power and energy, when it is either freed from the impediments which usually counteract it, or is itself of such force as entirely to repress those impediments. For as every body is susceptible of many different conditions, and has many different forms combined in it, one of them often confines, depresses, and hides another entirely, so that it is not easily distinguished. There are found, however, some subjects in which the *nature* inquired into is completely displayed, either by the absence of impediments, or by the predominance of its own power.

Bacon instances the thermometer, or *vitrum calendare*, as exhibiting the expansive power of heat, in a manner more distinct and measurable than in common cases. To this example, which is well chosen, the present state of science enables us to add many others.

If the weight of the air were inquired into, the Torricellian experiment or the barometer affords an *ostensive instance*, where the circumstance which conceals the weight of the atmosphere in common cases, namely, the pressure of it in all directions, being entirely removed, that weight produces its full effect, and sustains the whole column of mercury in the tube. The barometer affords also an example of the *instantia migrans*, when the change is not total, but only partial, or progressive. If it be the weight of the air which supports the mercury in the tube of the barometer, when that weight is diminished, the mercury ought to stand lower. On going to the top of a mountain, the weight of the incumbent air is diminished, because a shorter column of air is to be sustained; the mercury in the barometer ought therefore to sink, and it is found to do so accordingly.

These are instances in which the action of certain principles is rendered visible by the removal of all the opposing forces. One may be given where it is the distinct and decisive nature of the fact which leads to the result.

Suppose it were inquired, whether the present land had ever been covered by the sea. If we look at the stratified form of so large a portion of the earth's surface, we cannot but conclude it to be very probable that such land was formed at the bottom of the sea. But the decisive proof is afforded by the shells and corals, or bodies having the perfect shape of shells and corals, and of other marine exuviæ, which are found imbedded in masses of the most solid rock, and often on the tops of very high mountains. This leaves no doubt of the formation of the land under the sea, though it does not determine whether the land, since its formation, has been elevated to its present height, or the sea depressed to its present level. The decision of that question requires other facts to be consulted.

IV. The *instantia clandestina*, which is, as it were, opposed to the preceding, and shows some power or quality just as it is beginning to exist, and in its weakest state, is often very useful in the generalization of facts. Bacon also gave to this the fanciful name of *instantia crepusculi*.

An example of this may be given from hydrostatics. If the suspension of water in capillary tubes be inquired into, it becomes very useful to view that effect when it is least, or when the tube ceases to be capillary, and becomes a vessel of a large diameter. The column is then reduced to a slender ring of water which goes all round the vessel, and this, though now so inconsiderable, has the property of being independent of the size of the vessel, so as to be in all cases the same when the materials are the same. As there can be no doubt that this ring proceeds from the attraction of the sides, and of the part immediately above the water, so there can be no doubt that the capillary suspension, in part at least, is derived from the same cause. An effect of the opposite kind takes place when a glass vessel is filled with mercury.

V. Next to these may be placed what are called *instantiæ manipulares*, or collective instances, that is, general facts, or such as comprehend a great number of particular cases. As human knowledge can but seldom reach the most general cause or *form*, such collective instances are often the utmost extent to which our generalization can be carried. They have great value on this account, as they likewise have on account of the assistance which they give to farther generalization.

Of this we have a remarkable instance in one of the most important steps ever taken in any part of human knowledge. The laws of Kepler are facts of the kind now treated of, and consist of three general truths, each belonging to the whole planetary system, and it was by means of them that Newton discovered the principle of gravitation. The first is, that the planets all move in elliptical orbits, having the sun for their common focus; the next, that about this focus the *radius vector* of each planet describes equal areas in equal times. The third and last, that the squares of the periodic times of the planets are as the cubes of their mean distances from the sun. The knowledge of each of these was the result of much research, and of the comparison of a vast multitude of observations, insomuch that it may be doubted if ever three truths in science were discovered at the expence of so much labour and patience, or with the exertion of more ingenuity and invention in imagining and combining observations. These discoveries were all made before Bacon wrote, but he is silent concerning them; for the want of mathematical knowledge concealed from his view some of the most splendid and interesting parts of science.

Astronomy is full of such collective instances, and affords them, indeed, of the second and third order, that is to say, two or three times generalized. The astronomer observes nothing but that a certain luminous disk, or perhaps merely a luminous point, is in a certain position, in respect of the planes of the meridian and the horizon, at a certain moment of time. By comparing a number of such observations, he finds that this luminous point moves in a certain plane, with a certain velocity, and performs a revolution in a certain time. Thus, the periodic time of a planet is itself a collective fact, or a single fact expressing the result of many hundred observations. This holds with respect to each planet, and with respect to each element, as it is called, of the planet's orbit, every one of which is a general fact, expressing the result of an indefinite number of particulars. This holds still more remarkably of the inferences which extend to the distances of the planet from the earth, or from the sun. The laws of Kepler are therefore collective facts of the second, or even a higher order; or such as comprehend a great number of general facts, each of which is itself a general fact, including many particulars. It is much to the credit of astronomy, that, in all this process, no degree of truth or certainty is sacrificed; and that the same demonstrative evidence is preserved from the lowest to the highest point. Nothing but the use of mathematical reasoning could secure this advantage to any of the sciences.

VI. In the next place may be ranked the instances which Bacon calls *analogous*, or *parallel*. These consist of facts, between which an analogy or resemblance is visible in some particulars, notwithstanding great diversity in all the rest. Such are the telescope and microscope, in the works of art, compared with the eye in the works of nature. This, indeed, is an analogy which goes much beyond the mere exterior; it extends to the internal structure, and to the principle of action, which is the same in the eye and in the telescope,—to the *latent schematism*, in the language of Bacon, as far as material substance is concerned. It was the experiment of the *camera obscura* which led to the discovery of the formation of the images of external objects in the bottom of the eye by the action of the crystalline lens, and the other humours of which the eye is formed.

Among the instances of conformity, those are the most useful which enable us to compare productions of an unknown formation, with similar productions of which the formation is well understood. Such are basalt, and the other trap rocks, compared with the lava thrown out from volcanoes. They have a structure so exactly similar, that it is hardly possible to doubt that their origin is the same, and that they are both produced by the action of subterraneous fire. There are, however, amid their similarity, some very remarkable differences in the substances which they contain, the trap rocks containing calcareous spar, and the lava never containing any. On the supposition that they are both of igneous origin, is there any circumstance, in the conditions in which heat may have been applied to them, which can account for this difference? Sir James Hall, in a train of most philosophical and happily contrived experiments, has explained the nature of those conditions, and has shown that the presence of calcareous spar, or the want of it, may arise from the greater or less compression under which the fusion of the basalt was performed. This has served to explain a great difficulty in the history of the mineral kingdom.

Comparative anatomy is full of analogies of this kind, which are most instructive, and useful guides to discovery. It was by remarking in the blood-vessels a contrivance similar to the valves used in hydraulic engines, for preventing the counter current of a fluid, that Harvey was led to the discovery of the circulation of the blood. The analogies between natural and artificial productions are always highly deserving of notice.

The facts of this class, however, unless the analogy be very close, are apt to mislead, by representing accidental regularity as if it were constant. Of this we have an example in the supposed analogy between the colours in the prismatic spectrum and the divisions of the monochord. The colours in the prismatic spectrum do not occupy the same proportion of it in all cases: the analogy depends on the particular kind of glass, not on any thing that is common to all refraction. The tendency of man to find more order in things than there actually exists, is here to be cautiously watched over.

VII. *Monodic*, or singular facts, are the next in order. They comprehend the instances which are particularly distinguished from all those of the genus or species to which they belong. Such is the sun among the stars, the magnet among stones, mercury among metals, boiling fountains among springs, the elephant among quadrupeds. So also among the planets, Saturn is singular from his ring, the new planets are so likewise from their small size, from being extrazodiacal, &c.

Connected with these are the irregular and deviating instances, in which nature seems to depart from her ordinary course. Earthquakes, extraordinary tempests, years of great scarcity, winters of singular severity, &c. are of this number. All such facts ought to be carefully collected; and there should be added an account of all monstrous productions, and of every thing remarkable for its novelty and its rareness. Here, however, the most severe criticism must be applied; every thing connected with superstition is suspicious, as well as whatever relates to alchemy or magic.

A set of facts, which belongs to this class, consists of the instances in which *stones* have so often of late years been observed to fall from the heavens. Those stones are so unlike other atmospherical productions, and their origin must be so different from that of other minerals, that it is scarcely pos sible to imagine any thing more anomalous, and more inconsistent with the ordinary course of our experience. Yet the existence of this phenomenon is so well authenticated by testimony, and by the evidence arising from certain physical considerations, that no doubt with respect to it can be entertained, and it must therefore be received, as making a part of the natural history of meteors. But as every fact, or class of facts, which is perfectly singular, must be incapable of explanation, and can only be understood when its resemblance to other things has been discovered, so at present we are unable to assign the cause of these phenomena, and have no right to offer any theory of their origin.

VIII. Another class of facts is composed of what Bacon calls *instantiæ comitatus*, or examples of certain qualities which always accompany one another. Such are flame and heat,—flame being always accompanied by heat, and the same degree of heat in a given substance being always accompanied with flame. So also heat and expansion,—an increase of heat being accompanied with an increase of volume, except in a very few cases, and in circumstances very particular.

The most perfect *instantia comitatus* known, as being without any negative, is that of body and weight. Whatever is impenetrable and inert, is also heavy in a degree proportional to its inertia. To this there is no exception, though we do not perceive the connection as necessary.

Hostilen instances, or those of perpetual separation, are the reverse of the former.

Examples of this are found in air, and the other elastic fluids, which cannot have a solid form induced on them by any known means, when not combined with other substances. So also in solids, transparency and malleability are never joined, and appear to be incompatible, though it is not obvious for what reason.

IX. Passing over several classes which seem of inferior importance, we come to the *instantia crucis*, the division of this experimental logic which is most frequently resorted to in the practice of inductive investigation. When, in such an investigation, the understanding is placed in *equilibrio*, as it were, between two or more causes, each of which accounts equally well for the appearances, as far as they are known, nothing remains to be done but to look out for a fact which can be explained by the one of these causes, and not by the other; if such a one can be found, the uncertainty is removed, and the true cause is determined. Such facts perform the office of a cross, erected at the separation of two roads, to direct the traveller which he is to take, and, on this account, Bacon gave them the name of *instantiæ crucis*.

Suppose that the subject inquired into were the motion of the planets, and that the phenomena which first present themselves, or the motion of these bodies in longitude, could be explained equally on the Ptolemaic and the Copernican system, that is, either on the system which makes the Earth, or that which makes the Sun, the centre of the planetary motions, a cautious philosopher would hesitate about which of the two he should adopt, and notwithstanding that one of them was recommended by its superior simplicity, he might not think himself authorized to give to it a decided preference above the other. If, however, he consider the motion of these bodies in latitude, that is to say, their digressions from the plane of the ecliptic, he will find a set of phenomena which cannot be reconciled with the supposition that the earth is the centre of the planetary motions, but which receive the most simple and satisfactory explanation from supposing that the sun is at rest, and is the centre of those motions. The latter phenomena would therefore serve as *instantiæ crucis*, by which the superior credibility of the Copernican system was fully evinced.

Another example which I shall give of an *instantia crucis* is taken from chemistry, and is, indeed, one of the most remarkable experiments which has been made in that science.

It is a general fact observed in chemistry, that

metals are always rendered heavier by calcination. When a mass of tin or lead, for instance, is calcined in the fire, though every precaution is taken to prevent any addition from the adhesion of ashes, coals, &c. the absolute weight of the mass is always found to be increased. It was long before the cause of this phenomenon was understood. There might be some heavy substance added, though what it was could not easily be imagined; or some substance might have escaped, which was in its nature light, and possessed a tendency upwards. Other phenomena, into the nature of which it is at present unnecessary to inquire, induced chemists to suppose, that in calcination a certain substance actually escapes, being present in the regulus, but not in the calx of the metal. This substance, to which they gave the name of phlogiston, was probably that which, by its escape, rendered the metal heavier, and must, therefore, be itself endued with absolute levity.

The *instantia crucis* which extricated philosophers from this difficulty, was furnished by an experiment of the celebrated Lavoisier. That excellent chemist included a quantity of tin in a glass retort, hermetically sealed, and accurately weighed together with its contents; he then applied the necessary heat, and when the calcination of the tin was finished, he found the weight of the whole precisely the same as before. This proved, that no substance, which was either light or heavy, in a sensible degree, had made its way through the glass. The experiment went still farther. When the retort was cooled and opened, the air rushed in, so that it was evident that a part of the air had disappeared, or had lost its elasticity. On weigh ing the whole apparatus, it was now found that its weight was increased by ten grains; so that ten grains of air had entered into the retort when it was opened. The calx was next taken out, and weighed separately, and it was found to have become heavier by ten grains precisely. The ten grains of air then which had disappeared, and which had made way for the ten grains that rushed into the retort, had combined with the metal during the process of calcination. The farther prosecution of this very decisive experiment led to the knowledge of that species of air which combines with metals when they are calcined. The doctrine of phlogiston was of course exploded, and a creature of the imagination replaced by a real existence.

The principle which conducts to the contrivance of an *experimentum crucis* is not difficult to be understood. Taking either of the hypotheses, its consequences must be attempted to be traced, supposing a different experiment to be made. This must be done with respect to the other hypothesis, and a case will probably at last occur, where the two hypotheses would give different results. The experiment made in those circumstances will furnish an *instantia crucis*.

Thus, if the experiment of calcination be performed in a close vessel, and if phlogiston be the cause of the increase of weight, it must either escape through the vessel, or it must remain in the vessel after separation from the calx. If the former be the case, the apparatus will be increased in weight; if the latter, the phlogiston must make its escape on opening the vessel. If neither of these be the case, it is plain that the theory of phlogiston is insufficient to explain the facts.

The *experimentum crucis* is of such weight in matters of induction, that in all those branches of science where it cannot easily be resorted to, (the circumstances of an experiment being out of our power, and incapable of being varied at pleasure,) there is often a great want of conclusive evidence. This holds of agriculture, medicine, political economy, &c. To make one experiment, similar to another in all respects but one, is what the *experimentum crucis*, and, in general, the process of induction, principally requires; but it is what, in the sciences just named, can seldom be accomplished. Hence the great difficulty of separating the causes, and allotting to each its due proportion of the effect. Men deceive themselves in consequence of this continually, and think they are reasoning from fact and experience, when, in reality, they are only reasoning from a mixture of truth and falsehood. The only end answered by facts so incorrectly apprehended is that of making error more incorrigible.

Of the twenty-seven classes into which *instantiæ* are arranged by the author of the *Novum Organum*, fifteen immediately address themselves to the Understanding ; five serve to correct or to inform the Senses ; and seven to direct the hand in raising the superstructure of Art on the foundation of Science. The examples given above are from the first of these divisions, and will suffice for a summary. To the five that follow next, the general name of *instantiæ lampadis* is given, from their office of assisting or informing the senses.

Of these the *instantiæ januæ* assist the immediate action of the senses, and especially of sight. The examples quoted by Bacon are the microscope and telescope, (which last he mentions as the invention of Galileo,) and he speaks of them with great admiration, but with some doubt of their reality.

The *instantiæ citantes* enable us to perceive things which are in themselves insensible, or not at all the objects of perception. They cite or place things, as it were, before the bar of the senses, and from this analogy to judicial proceedings is derived the name of *instantiæ citantes*. Such, to employ examples which the progress of science has unfolded since the time of Bacon, are the air-pump and the barometer for manifesting the weight and elasticity of air; the measurement of the velocity of light, by means of the eclipses of the satellites of Jupiter, and the aberration of the fixed stars; the experiments in electricity and Galvanism, and in the greater part of pneumatic chemistry. In all these instances things are made known which before had entirely escaped the senses.

The *instantiæ viæ* are facts which manifest the continuous progress of nature in her operations. There is a propensity in men to view nature as it were at intervals, or at the ends of fixed periods, without regarding her gradual and unceasing action.* The desire of making observation easy is the great source of this propensity. Men wish for knowledge, but would obtain it at the least expence of time and labour. As there is no time, however, at which the hand of nature ceases to work, there ought to be none at which observation ceases to be made.

The *instantiæ persecantes*, or *vellicantes*, are those which force us to attend to things which, from their subtlety and minuteness, escape common observation.

Some of Bacon's remarks on this subtlety are such as would do credit to the most advanced state of science, and show how much his mind was fitted

* Nov. Org. II. Aph. 41.

for distinguishing and observing the great and admirable in the works of nature.

The last division contains seven classes, of which I mention only two. The experiments of this division are those most immediately tending to produce the improvement of art from the extension of science. "Now there are," says Bacon, "two ways in which knowledge, even when sound in itself, may fail of becoming a safe guide to the artist, and these are either when it is not sufficiently precise, or when it leads to more complicated means of producing an effect that can be employed in practice. There are therefore two kinds of experiments which are of great value in promoting the alliance between knowledge and power;—those which tend to give accurate and exact measures of objects, and those which disencumber the processes deduced from scientific principles of all unnecessary operations."

In the *instantiæ radii* we measure objects by lines and angles; in the *instantiæ curriculi* by time or by motion.

To the former of these classes are to be referred a number of instruments which now constitute the greater part of the apparatus of natural philosophy. Though Bacon had a just idea of their utility in general, he was unacquainted with most of them. The most remarkable at present are those that follow :

1. Astronomical instruments, or, more generally, all instruments for measuring lines and angles.

2. Instruments for measuring weight or force; such are the common balance, the hydrostatic balance, the barometer, the instruments used in England by Cavendish, and in France by Coulomb, which measure small and almost insensible actions by the force of torsion.

These last rather belong to the class of the *instantiæ luctæ*, where force is applied as the measure of force, than to the *instantiæ radii*.

3. The thermometer, newly invented in the time of Bacon, and mentioned by him under the name of *Vitrum Calendare*, an instrument to which we owe nearly all the knowledge we have of one of the most powerful agents in nature, viz. Heat.

4. The hygrometer, an instrument for measuring the quantity of humidity contained in the air; and in the construction of which, after repeated failures by the most skilful experimenters, the invention of Professor Leslie now promises success. Almost every one of these instruments, to which several more might be added, has brought in sight a new country, and has enriched science not only with new facts, but with new principles.

Among the remarks of Bacon on the *experimenta radii*, some are very remarkable for the extent of view which they display even in the infancy of physical science. He mentions the forces by which bodies act on one another at a distance, and throws out some hints at the attraction which the heavenly bodies exert on one another.

"Inquirendum est," says he, "si sit vis aliqua magnetica quæ operetur per consensum inter globum terræ et ponderosa, aut inter globum lunæ et aquas maris, aut inter cœlum stellatum et planetas, per quam evocentur et attollantur ad sua apogæa; hæc omnia operantur ad distantias admodum longinquas." *

Under the head of the *instantia curriculi*, or the measuring of things by time; after remarking that every change and every motion requires time, and illustrating this by a variety of instances, he has the following very curious anticipation of facts, which appeared then doubtful, but which subsequent discovery has ascertained:

"The consideration of those things produced in me a doubt altogether astonishing, viz. Whether the face of the serene and starry heavens be seen at the instant it really exists, or not till sometime later; and whether there be not, with respect to the heavenly bodies, a true time and an apparent time, no less than a true place and an apparent place, as astronomers say, on account of parallax. For it seems incredible that the *species* or rays of the celestial bodies can pass through the immense inter-

* Nov. Org. II. Aph. 45.

val between them and us in an instant, or that they do not even require some considerable portion of time." *

The measurement of the velocity of light, and the wonderful consequences arising from it, are the best commentaries on this passage, and the highest eulogy on its author.

Such were the speculations of Bacon, and the rules he laid down for the conduct of experimental inquiries, before any such inquiries had yet been instituted. The power and compass of a mind which could form such a plan beforehand, and trace not merely the outline, but many of the most minute ramifications, of sciences which did not yet exist, must be an object of admiration to all succeeding ages. He is destined, if, indeed, any thing in the world be so destined, to remain an *instantia singularis* among men, and as he has had no rival in the times which are past, so is he likely to have none in those which are to come. Before any parallel to him can be found, not only must a man of the same talents be produced, but he must be placed in the same circumstances; the memory of his predecessor must be effaced, and the light of science, after being entirely extinguished, must be again beginning to revive. If a second Bacon is ever to arise, he must be ignorant of the first.

---

* Nov. Org. II. Aph. 46.

Bacon is often compared with two great men who lived nearly about the same time with himself, and who were both eminent reformers of philosophy, Descartes and Galileo.

Descartes flourished about forty years later than Bacon, but does not seem to have been acquainted with his writings. Like him, however, he was forcibly struck with the defects of the ancient philosophy, and the total inaptitude of the methods which it followed, for all the purposes of physical investigation. Like him, too, he felt himself strongly impelled to undertake the reformation of this erroneous system ; but the resemblance between them goes no farther ; for it is impossible that two men could pursue the same end by methods more diametrically opposite.

Descartes never proposed to himself any thing which had the least resemblance to induction. He began with establishing principles, and from the existence of the Deity and his perfections, he proposed to deduce the explanation of all the phenomena of the world, by reasoning *a priori*. Instead of proceeding upward from the effect to the cause, he proceeded continually downward from the cause to the effect. It was in this manner that he sought to determine the laws of motion, and of the collision of bodies, in which last all his conclusions were erroneous. From the same source he deduced the existence of a *plenum*, and the continual preservation of the same quantity of motion in the universe; a proposition which, in a certain sense, is true, but in the sense in which he understood it, is altogether false. Reasonings of the kind which he employed may possibly suit, as Bacon observed, with intelligences of a higher order than man, but to his case they are quite inapplicable. Of the fruit of this tree nature has forbidden him to eat, and has ordained, that, with the sweat of his brow, and the labour of his hands, he should earn his knowledge as well as his subsistence.

Descartes, however, did not reject experiment altogether, though he assigned it a very subordinate place in his philosophy. By reasoning down from first principles, he tells us that he was always able to discover the effects; but the number of different shapes which those effects might assume was so great, that he could not determine, without having recourse to experiment, which of them nature had preferred to the rest. "We employ experiment," says he, "not as a reason by which any thing is proved, for we wish to deduce effects from their causes, and not conversely causes from their effects. We appeal to experience only, that out of innumerable effects which may be produced from the same cause, we may direct our attention to one rather than another." It is wonderful, that Descartes did not see what a severe censure he was here passing on himself; of how little value the speculations must be that led to conclusions so vague and indefinite; and how much more philosophy is disgraced by affording an explanation of things which *are not*, than by *not* affording an explanation of things which *are*.

As a system of philosophy and philosophic investigation, the method of Descartes can, therefore, stand in no comparison with that of Bacon. Yet his physics contributed to the advancement of science, but did so, much more by that which they demolished, than by that which they built up. In some particular branches the French philosopher far excelled the English. He greatly improved the science of optics, and in the pure mathematics, as has been already shown, he left behind him many marks of a great and original genius. He will, therefore, be always numbered among those who have essentially contributed to the advancement of knowledge, though nothing could be more perverse than his method of philosophizing, and nothing more likely to impede the progress of science, had not an impulse been at that time given to the human mind which nothing could resist.

Galileo, the other rival and contemporary of Bacon, is, in truth, one of those to whom human knowledge is under the greatest obligations. His discoveries in the theory of motion, in the laws of the descent of heavy bodies, and in the motion of projectiles, laid the foundation of all the great improvements which have since been made by the application of mathematics to natural philosophy. If to this we add the invention of the telescope, the discoveries made by that instrument, the confirmations of the Copernican system which these discoveries afforded, and, lastly, the wit and argument with which he combated and exposed the prejudice and presumption of the schools, we must admit that the history of human knowledge contains few greater names than that of Galileo. On comparing him with Bacon, however, I have no hesitation in saying, that the latter has given indications of a genius of a still higher order. In this I know that I differ from a historian, who was himself a philosopher, and whose suffrage, of consequence, is here of more than ordinary weight.

"The great glory of literature," says Hume, "in this island, during the reign of James, was Lord Bacon. If we consider the variety of talents displayed by this man, as a public speaker, a man of business, a wit, a courtier, a companion, an author, a philosopher, he is justly entitled to great admiration. If we consider him merely as an author and a philosopher, the light in which we view him at present, though very estimable, he was yet inferior to his contemporary Galileo, perhaps even to Kepler. Bacon pointed out, at a distance, the road to philosophy; Galileo both pointed it out to others, and made himself considerable advances in it. The Englishman was ignorant of geometry; the Florentine revived that science, excelled in it, and was the first who applied it, together with experiment, to natural philosophy. The former rejected, with the most positive disdain, the system of Copernicus; the latter fortified it with new proofs, derived both from reason and the senses. Bacon's style is stiff and rigid; his wit, though often brilliant, is also often unnatural and far-fetched. Galileo is a lively and agreeable, though somewhat a prolix writer." *

Though it cannot be denied that there is considerable truth in these remarks, yet it seems to me that the comparison is not made with the justness and discrimination which might have been expected from Hume, who appears studiously to have contrasted what is most excellent in Galileo, with what is most defective in Bacon. It is true that Galileo showed the way in the application of mathematics and of geometry to physical investigation, and that the immediate utility of his performance was greater than that of Bacon's; as it impressed more movement on the age in which he lived, example being always so much more powerful than precept. Bacon, indeed, wrote for an age more enlightened than his own, and it was

---

* Hist. of England, Vol. VI. Appendix.

long before the full merit of his work was understood. But though Galileo was a geometer, and Bacon unacquainted with the mathematics,—though Galileo added new proofs to the system of the earth's motion, which Bacon rejected altogether,—yet is it certain, I think, that the former has more fellows or equals in the world of science than the latter, and that his excellence, though so high, is less unrivalled. The range which Bacon's speculations embraced was altogether immense. He cast a penetrating eye on the whole of science, from its feeblest and most infantine state to that strength and perfection from which it was then so remote, and which it is perhaps destined to approach to continually, but never to attain. More substitutes might be found for Galileo than for Bacon. More than one could be mentioned, who, in the place of the former, would probably have done what he did; but the history of human knowledge points out nobody of whom it can be said, that, placed in the situation of Bacon, he would have done what Bacon did;—no man whose prophetic genius would have enabled him to delineate a system of science which had not yet begun to exist!—who could have derived the knowledge of what *ought to be* from what *was not*, and who could have become so rich in wisdom, though he received from his predecessors no inheritance but their errors. I am inclined, therefore, to agree with D'Alembert, "that when one considers the sound and enlarged views of this great man, the multitude of the objects to which his mind was turned, and the boldness of his style, which unites the most sublime images with the most rigorous precision, one is disposed to regard him as the greatest, the most universal, and the most eloquent of philosophers." *

3. Remarks, &c.

It will hardly be doubted by any one who attentively considers the method explained in the Novum Organum, which we have now attempted to sketch, that it contains a most comprehensive and rigorous plan of inductive investigation. A question, however, may occur, how far has this method been really carried into practice by those who have made the great discoveries in natural philosophy, and who have raised physical science to its present height in the scale of human knowledge? Is the whole method necessary, or have not circumstances occurred, which have rendered experimental investigation easier in practice than it appears to be in theory? To answer these questions completely, would require more discussion than is consistent with the limits of this Disserta-

* Discours Préliminaire de l'Encyclopédie.

tion; I shall, therefore, attempt no more than to point out the principles on which such an answer may be founded.

In a very extensive department of physical science, it cannot be doubted that investigation has been carried on, not perhaps more easily, but with a less frequent appeal to experience, than the rules of the Novum Organum would seem to require. In all the physical inquiries where mathematical reasoning has been employed, after a few principles have been established by experience, a vast multitude of truths, equally certain with the principles themselves, have been deduced from them by the mere application of geometry and algebra.

In mechanics, for example, after the laws of motion were discovered, which was done by experiment, the rest of the science, to a great extent, was carried on by reasoning from those laws, in the same manner that the geometer makes his discoveries by reasoning on the definitions, by help of a few axioms, or self-evident propositions. The only difference is, that, in the one case, the definitions and axioms are supplied solely from the mind itself, while, in the other, all the definitions and axioms, which are not those of pure geometry, are furnished by experience. *

* The part of mechanics which involves only statical con-

Bacon certainly was not fully aware of the advantages that were thus to accrue to the physical sciences. He was not ignorant, that the introduction of mathematical reasoning into those sciences is not only possible, but that, under certain conditions, it may be attended with the greatest advantage. He knew also in what manner this application had been abused by the Platonists, who had attempted, by means of geometry, to establish the first principles of physics, or had used them, *in axiomatis constituendis*, which is exactly the province belonging exclusively to experience. At the same time, he pointed out, with great precision, the place which the mathematics may legitimately occupy, as serving to measure and compare the objects of physical inquiry. He did not, however, perceive beforehand, nor was it possible that he should, the vast extent to which the application of that science was capable of being carried. In the book, De Augmentis, he has made many excellent remarks on this subject, full of the sagacity which penetrated so far into futurity, but, nevertheless, could only perceive a small part of the scene which the genius of Newton was afterwards to unfold.

---

siderations, or the equilibrium of forces, is capable of being treated by *reasoning a priori* entirely, without any appeal to experience. This will appear, when the subject of Mechanics is more particularly treated of.

Hence, the route which leads to many of the richest and most fertile fields of science, is not precisely that which Bacon pointed out; it is safer and easier, so that the voyager finds he can trust to his chart and compass alone, without constantly looking out, or having the sounding-line perpetually in his hand.

Another remark I must make on Bacon's method is, that it does not give sufficient importance to the *instantiæ radii*, or those which furnish us with accurate measures of physical quantities. The experiments of this class are introduced as only subservient to practice; they are, however, of infinite value in the theoretical part of induction, or for ascertaining the causes and essences of the things inquired into. We have an instance of this in the discovery of that important truth in physical astronomy, that the moon is retained in her orbit by the force of gravity, or the same which, at the earth's surface, makes a stone fall to the ground. This proposition, however it might have been suspected to be true, could never have been demonstrated but by such observations and experiments as assigned accurate geometrical measures to the quantities compared. The semidiameter of the earth; the velocity of falling bodies at the earth's surface; the distance of the moon, and her velocity in her orbit;—all these four elements must be determined with great precision, and afterwards compared together by certain theorems deduced from the laws of motion, before the relation between the force which retains the moon in her orbit, and that which draws a stone to the ground, could possibly be discovered. The discovery also, when made, carried with it the evidence of demonstration, so that here, as in many other cases, the *instantiæ radii* are of the utmost importance in the theoretical part of physics.

Another thing to be observed is, that, in many cases, the result of a number of particular facts, or the collective instance arising from them, can only be found out by geometry, which, therefore, becomes a necessary instrument in completing the work of induction. An example, which the science of optics furnishes, will make this clearer than any general description. When light passes from one transparent medium to another it is refracted, that is, it ceases to go on in a straight line, and the angle which the incident ray makes with the superficies which bounds the two *media*, determines that which the refracted ray makes with the same superficies. Now, if we would learn any thing about the relation which these angles bear to one another, we must have recourse to experiment, and all that experiment can do is, for any particular angle of incidence, to determine the corresponding angle of refraction. This may be done in innumerable cases; but, with respect to the general rule which, in every possible case, determines the one of those angles from the other, or expresses the constant and invariable relation which subsists between them, —with respect to it, experiment gives no direct information. The methods of geometry must therefore be called in to our assistance, which, when a constant though unknown relation subsists between two angles, or two variable quantities of any kind, and when an indefinite number of values of those quantities are given, furnishes infallible means of discovering that unknown relation, either accurately, or at least by approximation. In this way it has been found, that, when the two *media* remain the same, the cosines of the angles above mentioned have a constant ratio to one another. Thus it appears, that, after experiment has done its utmost, geometry must be applied before the business of induction can be completed. This can only happen when the experiments afford accurate measures of the quantities concerned, like the *instantiæ radii, curriculi*, &c. and this advantage of admitting generalization with so much certainty is one of their properties, of which it does not appear that even Bacon himself was aware.

Again, from the intimate connection which prevails among the principles of science, the success of one investigation must often contribute to the success of another, in such a degree as to make it unnecessary to employ the complete apparatus of induction from the beginning. When certain leading principles have been once established, they serve, in new investigations, to narrow the limits within which the thing sought for is contained, and enable the inquirer to arrive more speedily at the truth.

Thus, suppose that, after the nature of the reflection and refraction of light, and particularly of the colours produced by the latter, had been discovered by experiment, the cause of the rainbow were to be inquired into. It would, after a little consideration, appear probable, that the phenomenon to be explained depends on the reflection and refraction of light by the rain falling from a cloud opposite to the sun. Now, since the nature of reflection and refraction are supposed known, we have the principles previously ascertained which are likely to assist in the explanation of the rainbow. We have no occasion, therefore, to enter on the inquiry, as if the powers to be investigated were wholly unknown. It is the combination of them only which is unknown, and our business is to seek so to combine them, that the result may correspond with the appearances. This last is precisely what Newton accomplished, when, by deducing from the known laws of refraction and reflection the breadth of the coloured arch, the diameter of the circle of which it is a part, and the relation of the latter to the place of the spectator and of the sun, he found all these to come out from his calculus, just as they are observed in nature. Thus he proved the truth of his solution by the most clear and irresistible evidence.

The strict method of Bacon is therefore only necessary where the thing to be explained is new, and where we have no knowledge, or next to none, of the powers employed. This is but rarely the case, at least in some of the branches of Physics; and, therefore, it occurs most commonly in actual investigation, that the inquirer finds himself limited, almost from the first outset, to two or three hypotheses, all other suppositions involving inconsistencies which cannot for a moment be admitted. His business, therefore, is to compare the results of these hypotheses, and to consider what consequences may in any case arise from the one that would not arise from the other. If any such difference can be found, and if the matter is a subject of experiment, we have then an *instantia crucis* which must decide the question.

Thus, the *instantia crucis* comes in real practice to be the experiment most frequently appealed to, and that from which the most valuable information is derived.

In executing the method here referred to, the application of much reasoning, and frequently of much mathematical reasoning, is necessary, before any appeal to the experiment can be made, in order to deduce from each of the hypotheses an exact estimate of the consequences to which it leads. Suppose, for instance, that the law by which the magnetic virtue decreases in its intensity, as we recede from its poles, were to be inquired into. It is obvious that the number of hypotheses is here indefinite; and that we have hardly any choice but to begin with the simplest, or with that which is most analogous to the law of other forces propagated from a centre. Whatever law we assume, we must enter into a good deal of geometric reasoning, before a conclusion can be obtained, capable of being brought to the test of experience. The force itself, like all other forces, is not directly perceived; and its effects are not the result of its mere intensity, but of that intensity combined with the figure and magnitude of the body on which it acts; and, therefore, the calculus must be employed to express the measure of the effect, in terms of the intensity and the distance only. This being done, the hypothesis which gives results most nearly corresponding to the facts observed, when the magnet acts on the same body, at different distances, must be taken as the nearest approximation to the truth. We have here an instance of the use of hypothesis in inductive investigation, and, indeed, of the only legitimate use to which it can ever be applied.

It also appears that Bacon placed the ultimate object of philosophy too high, and too much out of the reach of man, even when his exertions are most skilfully conducted. He seems to have thought, that, by giving a proper direction to our researches, and carrying them on according to the inductive method, we should arrive at the knowledge of the essences of the powers and qualities residing in bodies; that we should, for instance, become acquainted with the essence of heat, of cold, of colour, of transparency. The fact, however, is, that, in as far as science has yet advanced, no one essence has been discovered, either as to matter in general, or as to any of its more extensive modifications. We are yet in doubt, whether heat is a peculiar motion of the minute parts of bodies, as Bacon himself conceived it to be; or something emitted or radiated from their surfaces; or lastly, the vibrations of an elastic medium, by which they are penetrated and surrounded. Yet whatever be the form or essence of heat, we have discovered a great number of its properties and its laws; and have done so, by pursuing with more or less accuracy the method of induction. We have also this consolation for the imperfection of our theoretical knowledge, that, in as much as art is concerned, or the possession of power over heat, we have perhaps all the advantages that could be obtained from a complete knowledge of its essence.

An equal degree of mystery hangs over the other properties and modifications of body; light, electricity, magnetism, elasticity, gravity, are all in the same circumstances; and the only advance that philosophy has made toward the discovery of the essences of these qualities or substances is, by exploding some theories, rather than by establishing any,—so true is Bacon's maxim, that the first steps in philosophy necessarily consist in negative propositions. Besides this, in all the above instances the laws of action have been ascertained; the phenomena have been reduced to a few general facts, and in some cases, as in that of gravity, to one only; and for ought that yet appears, this is the highest point which our science is destined to reach.

In consequence of supposing a greater perfection in knowledge than is ever likely to be attained, Bacon appears, in some respects, to have misapprehended the way in which it is ultimately to become applicable to art. He conceives that, if the *form* of any quality were known, we should be able, by inducing that form on any body, to communicate to it the said quality. It is not probable, however, that this would often lead to a more easy and simple process than that which art has already invented. In the case of colour, for example, though ignorant of its *form*, or of the construction of surface which enables bodies to reflect only light of a particular species, yet we know how to communicate that power from one body to another. Nor is it likely, though this structure were known with ever so great precision, that we should be able to impart it to bodies by any means so simple and easy, as by the common process of immersing them in a liquid of a given colour.

In some instances, however, the theories of chemistry have led to improvements of art very conformable to the anticipations of the *Novum Organum*. A remarkable instance of this occurs in the process for bleaching, invented by Berthollet. It had been for some time known, that the combination of the chemical principle of oxygen with the colouring matter in bodies, destroyed, or discharged, the colour ; and that, in the common process of bleaching, it was chiefly by the union of the oxygen of the air with the colouring matter in the cloth that this effect was produced. The excellent chemist just named conceived, therefore, that if the oxygen could be presented to the cloth in a dense state, and, at the same time, feebly combined with any other body, it might unite itself to the colouring matter so readily, that the process of bleaching would by that means be greatly accelerated. His skill in chemistry suggested to him a way in which this might easily be done, by immersing the cloth in a liquid containing much oxygen in a loose state, or one in which it was slightly combined with other substances, and the effect followed so exactly, that he was able to perform in a few hours what required weeks, and even months, according to the common process.

This improvement, therefore, was a real gift from the sciences to the arts; and came nearly, though not altogether, up to the ideas of Bacon. I suspect not altogether, because the manner in which oxygen destroys the colour of bodies, or alters the structure of their surfaces, remains quite unknown.

It was natural, however, that Bacon, who studied these subjects theoretically, and saw no-where any practical result in which he could confide, should listen to the inspirations of his own genius, and ascribe to philosophy a perfection which it may be destined never to attain. He knew, that from what it had not yet done, he could conclude nothing against what it might hereafter accomplish. But after his method has been followed, as it has now been, with greater or less accuracy, for more than two hundred years, circumstances are greatly changed; and the impediments, which, during all that time, have not yielded in the least to any effort, are perhaps never likely to be removed. This may, however, be a rash inference; Bacon, after all, may be in the right; and we may be judging under the influence of the vulgar prejudice, which has convinced men, in every age, that they had nearly reached the farthest verge of human knowledge. This must be left for the decision of posterity; and we should rejoice to think, that judgment will hereafter be given against the opinion which at this moment appears most probable.

SECTION III.

MECHANICS.

1. THEORY OF MOTION.

BEFORE the end of the sixteenth century, mechanical science had never gone beyond the problems which treat of the equilibrium of bodies, and had been able to resolve these accurately, only in the cases which can be easily reduced to the lever. Guido Ubaldi, an Italian mathematician, was among the first who attempted to go farther than Archimedes and the ancients had done in such inquiries. In a treatise which bears the date of 1577, he reduced the pulley to the lever, but with respect to the inclined plane, he continued in the same error with Pappus Alexandrinus, supposing that a certain force must be applied to sustain a body, even on a plane which has no inclination.

Stevinus, an engineer of the Low Countries, is the first who can be said to have passed beyond the point at which the ancients had stopped, by determining accurately the force necessary to sustain a body on a plane inclined at any angle to the horizon. He resolved also a great number of other problems connected with the preceding, but, nevertheless, did not discover the general principle of the composition of forces, though he became acquainted with this particular case, immediately applicable to the inclined plane.

The remark, that a chain laid on an inclined plane, with a part of it hanging over at top, in a perpendicular line, will be *in equilibrio*, if the two ends of the chain reach down exactly to the same level, led him to the conclusion, that a body may be supported on such a plane by a force which draws in a direction parallel to it, and has to the weight of the body the same ratio that the height of the plane has to its length.

Though it was probably from experience that Stevinus derived the knowledge of this proposition, he attempted to prove the truth of it by reasoning *a priori*. He supposed the two extremities of the chain, when disposed as above, to be connected by a part similar to the rest, which, therefore, must hang down, and form an arch. If in this state, says he, the chain were to move at all, it would continue to move for ever, because its situation, on the whole, never changing, if it were determined to move at one instant, it must be so determined at every other instant. Now, such perpetual motion, he adds, is impossible, and therefore the chain, as here supposed, with the arch hanging below, does not move. But the force of the arch below draws down the ends of the chain equally, because the arch is divided in the middle or lowest point into two parts similar and equal. Take away these two equal forces, and the remaining forces will also be equal, that is, the tendency of the chain to descend along the inclined plane, and the opposite tendency of the part hanging perpendicularly down, are equal, or are *in equilibrio* with one another. Such is the reasoning of Stevinus, which, whether perfectly satisfactory or not, must be acknowledged to be extremely ingenious, and highly deserving of attention, as having furnished the first solution of a problem, by which the progress of mechanical science had been long arrested. The first appearance of his solution is said to have had the date of 1585; but his works, as we now see them, were collected after his death, by his countryman Albert Girard, and published at Leyden in 1634. * Some discoveries of Stevinus in hydrostatics will be hereafter mentioned.

The person who comes next in the history of mechanics made a great revolution in the physical sciences. Galileo was born at Pisa in the year

* The edition of Albert Girard is entitled *Œuvres Mathématiques de Stevins,* in folio. See Livre I. *De la Statique,* Theorem 11. Stevinus also wrote a treatise on navigation, which was published in Flemish in 1586, and was afterwards honoured with a translation into Latin, by Grotius. The merit of Stevinus has been particularly noticed by La Grange. *Mécanique Analytique,* Tom. I. Sect. 1. § 5.

1564. He early applied himself to the study of mathematics and natural philosophy; and it is from the period of his discoveries that we are to date the joint application of experimental and geometrical reasoning to explain the phenomena of nature.

As early as 1592 he published a treatise, *della Scienza Mecanica*, in which he has given the theory, not of the lever only, but of the inclined plane and the screw; and has also laid down this general proposition, that mechanical engines make a small force equivalent to a great one, by making the former move over a greater space in the same time than the latter, just in proportion as it is less. No contrivance can make a small weight put a great one in motion, but such a one as gives to the small weight a velocity which is as much greater than that of the large weight, as this last weight is greater than the first. These general propositions, and their influence on the action of machinery, Galileo proceeded to illustrate with that clearness, simplicity, and extent of view, in which he was quite unrivalled; and hence, I think, it is fair to consider him as the first person to whom the mechanical principle, since denominated that of *the virtual velocities*, had occurred in its full extent. The object of his consideration was the action of machines in motion, and not merely of machines *in equilibrio*, or at rest; and he showed, that, if the effect of a force be estimated by the weight it can raise to a given beight in a given time, this effect can never be increased by any mechanical contrivance whatsoever.

In the same treatise, he lays it down as a postulate, (*supposizione,*) that the effect of one heavy body to turn another round a centre of motion, is proportional to the perpendicular drawn from that centre to the vertical passing through the body, or in general to the direction of the force. This proposition he states without a demonstration, and passes by means of it to the oblique lever, and thence to the inclined plane. To speak strictly, however, the demonstrations with respect to both these last are incomplete, the preceding proposition being assumed in them without proof. It is probable that he satisfied himself of the truth of it, on the principle, that the distances of forces from the centre of motion must always be measured by lines making the *same* angles with their directions, and that of such lines the simplest are the perpendiculars. His demonstration is regarded by La Grange as quite satisfactory. *

Galileo extended the theory of motion still farther. He had begun, while pursuing his studies at the university of Pisa, to make experiments on the descent of falling bodies, and discovered the fact, that heavy and light bodies fall to the ground from the same height in the same time, or in times

* Mécanique Analytique, Tom I. Sect. 1. § 6.

so nearly the same, that the difference can only be ascribed to the resistance of the air. From observing the vibrations of the lamps in the cathedral, he also arrived at this very important conclusion in mechanics, that the great and the small vibrations of the same pendulum are performed in the same time, and that this time depends only on the length of the pendulum. The date of these observations goes back as far as the year 1583.

These experiments drew upon him the displeasure of his masters, who considered it as unbecoming of their pupil to seek for truth in the Book of Nature rather than in the writings of Aristotle, when elucidated by their commentaries, and from that moment began the persecutions with which the prejudice, the jealousy, and bigotry of his contemporaries continued to harass or afflict this great man throughout his whole life.

That the acceleration of falling bodies is uniform, or, that they receive equal increments of velocity in equal times, he appears first to have assumed as the law which they follow, merely on account of its simplicity. Having once assumed this principle, he showed, by mathematical reasoning, that the spaces descended through must be as the squares of the times, and that the space fallen through in one second is just the half of that which the body would have described in the same time with the velocity last acquired.

The knowledge which he already had of the properties of the inclined plane enabled him very readily to perceive, that a body descending on such a plane must be uniformly accelerated, though more slowly than when it falls directly, and is accelerated by its whole weight. By means of the inclined plane, therefore, he was able to bring the whole theory of falling bodies to the test of experiment, and to prove the truth of his original assumption, the uniformity of their acceleration.

His next step was to determine the path of a heavy body, when obliquely projected. He showed this path to be a parabola; and here, for the first time, occurs the use of a principle which is the same with the composition of motion in its full extent. Galileo, however, gave no name to this principle; he did not enunciate it generally, nor did he give any demonstration of it, though he employed it in his reasonings. The inertia of body was assumed in the same manner; it was, indeed, involved in the uniform acceleration of falling bodies, for these bodies did not lose in one minute the motion acquired in the preceding, but, retaining it, went on continually receiving more.

The theory of the inclined plane had led to the knowledge of this proposition, that, if a circle be placed vertically, the chords of different arches terminating in the lowest point of the circle, are all descended through in the same space of time.

This seemed to explain why, in a circle, the great and the small vibrations are of equal duration. Here, however, Galileo was under a mistake, as the motions in the chord and in the arch are very dissimilar. The accelerating force in the chord remains the same from the beginning to the end, but, in the arch, it varies continually, and becomes, at the lowest point, equal to nothing. The times in the chords, and in the arches, are therefore different, so that here we have a point marking the greatest distance in this quarter, to which the mechanical discoveries of Galileo extended. The first person who investigated the exact time of a vibration in an arch of a circle was Huygens, a very profound mathematician.

To this list of mechanical discoveries, already so important and extensive, we must add, that Galileo was the first who maintained the existence of the *law of continuity*, and who made use of it as a principle in his reasonings on the phenomena of motion. *

The vibrations of the pendulum having suggested to Galileo the means of measuring time accurately, it appears certain that the idea of applying it to the clock had also occurred to him, and of using the chronometer so formed for finding the

---

* Opere di Galileo, Tom. IV. Dial. 1. p. 32, Florence edition, and in many other parts.

longitude, by means of observations made on the eclipses of the satellites of Jupiter. How far he had actually proceeded in an invention which required great practical knowledge, and which afterwards did so much credit to Huygens, appears to be uncertain, and not now easy to be ascertained. But that the project had occurred to him, and that he had taken some steps towards realizing it, is sufficiently established.

One forms, however, a very imperfect idea of this philosopher, from considering the discoveries and inventions, numerous and splendid as they are, of which he was the undisputed author. It is by following his reasonings, and by pursuing the train of his thoughts in his own elegant, though somewhat diffuse exposition of them, that we become acquainted with the fertility of his genius, with the sagacity, penetration, and comprehensiveness of his mind. The service which he rendered to real knowledge is to be estimated not only from the truths which he discovered, but from the errors which he detected,—not merely from the sound principles which he established, but from the pernicious idols which he overthrew. His acuteness was strongly displayed in the address with which he exposed the errors of his adversaries, and refuted their opinions, by comparing one part of them with another, and proving their extreme inconsistency. Of all the writers who have lived in an age, which was yet only emerging from ignorance and barbarism, Galileo has most entirely the tone of true philosophy, and is most free from any contamination of the times in taste, sentiment, and opinion.

The discoveries of this great man concerning motion drew the attention of philosophers more readily, from the circumstance that the astronomical theories of Copernicus had directed their attention to the same subject. It had become evident, that the great point in dispute between his system and the Ptolemaic must be finally decided by an appeal to the nature of motion and its laws. The great argument to which the friends of the latter system naturally had recourse was the impossibility, as it seemed to them to be, of the swift motion of the earth being able to exist, without the perception, nay, even without the destruction, of its inhabitants. It was natural for the followers of Copernicus to reply, that it was not certain that these two things were incompatible; that there were many cases in which it appeared, that the motion common to a whole system of bodies did not affect the motion of those bodies relatively to one another; that the question must be more deeply inquired into; and that, without this, the evidence on opposite sides could not be fairly and accurately compared. Thus it was, at a very fortunate moment, that Galileo made his discoveries in Mechanics, as they were rendered more interesting by those which, at that very time, he himself was making in Astronomy. The system of Copernicus had, in this manner, an influence on the theory of motion, and, of course, on all the parts of natural philosophy. The *inertia* of matter, or, the tendency of body, when put in motion, to preserve the quantity and direction of that motion, after the cause which impressed it has ceased to act, is a principle which might still have been unknown, if it had not been forced upon us by the discovery of the motion of the earth.

The first addition which was made to the mechanical discoveries of Galileo was by Torricelli, in a treatise *De Motu Gravium naturaliter descendentium et projectorum.* * To this ingenious man we are indebted for the discovery of a remarkable property of the centre of gravity, and a general principle with respect to the equilibrium of bodies. It is this: If there be any number of heavy bodies connected together, and so circumstanced, that by their motion their centre of gravity can neither ascend nor descend, these bodies will remain at rest. This proposition often furnishes the means of resolving very difficult questions in mechanics.

Descartes, whose name is so great in philosophy and mathematics, has also a place in the history of

* Vitæ Italorum Illustrium, Vol. I. p. 347.

mechanical discovery. With regard to the action of machines, he laid down the same principle which Galileo had established,—that an equal effort is necessary to give to a weight a certain velocity, as to give to double the weight the half of that velocity, and so on in proportion, the effect being always measured by the weight multiplied into the velocity which it receives. He could hardly be ignorant that this proposition had been already stated by Galileo, but he has made no mention of it. He, indeed, always affected a disrespect for the reasonings and opinions of the Italian philosopher, which has done him no credit in the eyes of posterity.

The theory of motion, however, has in some points been considerably indebted to Descartes. Though the reasonings of Galileo certainly involve the knowledge of the disposition which matter has to preserve its condition either of rest or of rectilineal and uniform motion, the first distinct enunciation of this law is found in the writings of the French philosopher. It is, however, there represented, not as mere inactivity, or indifference, but as a real force, which bodies exert in order to preserve their state of rest or of motion, and this inaccuracy affects some of the reasonings concerning their action on one another.

Descartes, however, argued very justly, that all motion being naturally rectilineal, when a body moves in a curve, this must arise from some constraint, or some force urging it in a direction different from that of the first impulse, and that if this cause were removed at any time, the motion would become rectilineal, and would be in the direction of a tangent, to the curve at the point where the deflecting force ceased to act.

Lastly, He taught that the quantity of motion in the universe remains always the same.

The reasoning by which he supported the first and second of these propositions is not very convincing, and though he might have appealed to experience for the truth of both, it was not in the spirit of his philosophy to take that method of demonstrating its principles. His argument was, that motion is a state of body, and that body or matter cannot change its own state. This was his demonstration of the first proposition, from which the second followed necessarily.

The evidence produced for the third, or the preservation of the same quantity of motion in the universe, is founded on the immutability of the Divine nature, and is an instance of the intolerable presumption which so often distinguished the reasonings of this philosopher. Though the immutability of the Divine nature will readily be admitted, it remains to be shown, that the continuance of the same quantity of motion in the universe is a consequence of it. This, indeed, cannot be shown, for that quantity, in the sense in which Descartes understood it, is so far from being preserved uniform, that it varies continually from one instant to another. It is nevertheless true, that the quantity of motion in the universe, when rightly estimated, is invariable, that is, when reduced to the direction of three axes at right angles to one another, and when opposite motions are supposed to have opposite signs. This is a truth now perfectly understood, and is a corollary to the equality of action and reaction, in consequence of which, whatever motion is communicated in one direction, is either lost in that direction, or generated in the opposite. This, however, is quite different from the proposition of Descartes, and if expressed in his language, would assert, not that the sum, but that the difference of the opposite motions in the universe remains constantly the same. When he proceeds, by help of the principle which he had thus mistaken, to determine the laws of the collision of bodies, his conclusions are almost all false, and have, indeed, such a want of consistency and analogy with one another, as ought, in the eyes of a mathematician, to have appeared the most decisive indications of error. How this escaped the penetration of a man well acquainted with the harmony of geometrical truths, and the gradual transitions by which they always pass into one another, is not easily explained, and, perhaps, of all his errors, is the least consistent with the powerful and systematic genius which he is so well known to have possessed.

Thus, the obligation which the theory of motion has to this philosopher, consists in his having pointed out the nature of centrifugal force, and ascribed that force to the true cause, the inertia of body, or its tendency to uniform and rectilineal motion.

The laws which actually regulate the collision of bodies remained unknown till some years later, when they were recommended by the Royal Society of London to the particular attention of its members. Three papers soon appeared, in which these laws were all correctly laid down, though no one of the authors had any knowledge of the conclusions obtained by the other two. The first of these was read to the Society, in November 1668, by Dr Wallis of Oxford; the next by Sir Christopher Wren, in the month following, and the third by Huygens, in January 1669. The equality of action and reaction, and the maxim, that the same force communicates to different bodies velocities which are inversely as their masses, are the principles on which these investigations are founded.

The ingenious and profound mathematician last mentioned is also the first who explained the true relation between the length of a pendulum, and the time of its least vibrations, and gave a rule by which the time of the rectilineal descent, through a line equal in length to the pendulum, might from thence be deduced. He next applied the pendulum to regulate the motion of a clock, and gave an account of his construction, and the principles of it, in his *Horologium Oscillatorium*, about the year 1670, though the date of the invention goes as far back as 1656.* Lastly, He taught how to correct the imperfection of a pendulum, by making it vibrate between cycloidal cheeks, in consequence of which its vibrations, whether great or small, became, not approximately, but precisely of equal duration.

Robert Hooke, a very celebrated English mechanician, laid claim to the same application of the pendulum to the clock, and the same use of the cycloidal cheeks. There is, however, no dispute as to the priority of Huygens' claim, the invention of Hooke being as late as 1670. Of the cycloidal cheeks, he is not likely to have been even the second inventor. Experiment could hardly lead any one to this discovery, and he was not sufficiently skilled in the mathematics to have found it out by mere reasoning. The fact is, that though very original and inventive, Hooke was jealous and illiberal in the extreme; he appropriated to himself the inventions of all the world, and accused all the world of appropriating his.

---

* Montucla, Tom. II. p. 418, 2d edit.

It has already been observed, that Galileo conceived the application of the pendulum to the clock earlier, by several years, than either of the periods just referred to. The invention did great honour to him and to his two rivals; but that which argues the most profound thinker, and the most skilful mathematician of the three, is the discovery of the relation between the length of the pendulum and the time of its vibration, and this discovery belongs exclusively to Huygens. The method which he followed in his investigation, availing himself of the properties of the cycloid, though it be circuitous, is ingenious, and highly instructive.

An invention, in which Hooke has certainly the priority to any one, is the application of a spiral spring to regulate the balance of a watch. It is well known of what practical utility this invention has been found, and how much it has contributed to the solution of the problem of finding the longitude at sea, to which not only he, but Galileo and Huygens, appear all to have had an eye.

In what respects the theory of motion, Huygens has still another strong claim to our notice. This arises from his solution of the problem of finding the centre of oscillation of a compound pendulum, or the length of the simple pendulum vibrating in the same time with it. Without the solution of this problem, the conclusions respecting the pendulum were inapplicable to the construction of clocks, in which the pendulums used are of necessity compound. The problem was by no means easy, and Huygens was obliged to introduce a principle which had not before been recognised, that if the compound pendulum, after descending to its lowest point, was to be separated into particles distinct and unconnected with one another, and each left at liberty to continue its own vibration, the common centre of gravity of all those detached weights would ascend to the same height to which it would have ascended had they continued to constitute one body. The above principle led him to the true solution, and his investigation, though less satisfactory than those which have been since given, does great credit to his ingenuity. This was the most difficult mechanical inquiry which preceded the invention of the differential or fluxionary calculus.

2. Hydrostatics.

While the theory of motion, as applied to solids, was thus extended, in what related to fluids, it was making equal progress. The laws which determine the weight of bodies immersed in fluids, and also the position of bodies floating on them, had been discovered by Archimedes, and were farther illustrated by Galileo. It had also been discovered by Stevinus, that the pressure of fluids is in proportion to their depth, and thus the two leading principles of hydrostatics were established. Hydraulics, or the motion of fluids, was a matter of more difficulty, and here the first step is to be ascribed to Torricelli, who, though younger than Galileo, was for some time his contemporary. He proved that water issues from a hole in the side or bottom of a vessel, with the velocity which a body would acquire, by falling from the level of the surface to the level of the orifice. This proposition, now so well known as the basis of the whole doctrine of Hydraulics, was first published by Torricelli at the end of his book, *De Motu Gravium et Projectorum ;* but it is not the greatest discovery which science owes to the friend and disciple of Galileo. The latter had failed in assigning the reason why water cannot be raised in pumps higher than thirty-three feet, but he had remarked, that if a pump is more than thirty-three feet in length, a vacuum will be left in it. Torricelli, reflecting on this, conceived, that if a heavier fluid than water were used, a vacuum might be produced, in a way far shorter, and more compendious. He tried mercury, therefore, and made use of a glass tube about three feet long, open at one end, and close at the other, where it terminated in a globe. He filled this tube, shut it with his finger, and inverted it in a basin of mercury. The result is well known ;

—he found that a column of mercury was suspended in the tube, an effect which he immediately ascribed to the pressure of the atmosphere. So disinterested was this philosopher, however, that he is said to have lamented that Galileo, when inquiring into the cause why water does not ascend in pumps above a certain height, had not discovered the true cause of the phenomenon. The generosity of Torricelli was perhaps rarer than his genius; —there are more who might have discovered the suspension of mercury in the barometer, than who would have been willing to part with the honour of the discovery to a master or a friend.

This experiment opened the door to a multitude of new discoveries, and demolished a formidable idol, the horror of a vacuum, to which so much power had been long attributed, and before which even Galileo himself had condescended to bow.

The objections which were made to the explanation of the suspension of the mercury in the tube of the barometer, were overthrown by carrying that instrument to the top of Puy de Dôme, an experiment suggested by Pascal. The descent of the mercury showed, that the pressure which supported it was less there than at the bottom; and it was afterwards found, that the fall of the mercury corresponded exactly to the diminution of the length of the pressing column, so that it afforded a measure of that diminution, and, consequently, of the heights of mountains. The invention of the air-pump by Otto Guericke, burgomaster of Magdeburg, quickly followed that of the barometer by Torricelli, though it does not appear that the invention of the Italian philosopher was known to the German. In order to obtain a space entirely void of air, Otto Guericke filled a barrel with water, and having closed it exactly on all sides, began to draw out the water by a sucking-pump applied to the lower part of the vessel. He had proceeded but a very little way, when the air burst into the barrel with a loud noise, and its weight was proved by the failure of the experiment, as effectually as it could have been by its success. After some other trials, which also failed, he thought of employing a sphere of glass, when the experiment succeeded, and a vacuum was obtained. This was about the year 1654.

The elasticity of the air, as well as its weight, now became known; its necessity to combustion, and the absorption of a certain proportion of it, during that process; its necessity for conveying sound;—all these things were clearly demonstrated. The necessity of air to the respiration of animals required no proof from experiment, but the sudden extinction of life, by immersion in a vacuum, was a new illustration of the fact.

The first considerable improvements made on the air-pump are due to Mr Boyle. He substituted to the glass globe of Otto Guericke a receiver of a more commodious form, and constructed his pump so as to be worked with much more facility. His experiments were farther extended,—they placed the weight and elasticity of the air in a variety of new lights,—they made known the power of air to dissolve water, &c. Boyle had great skill in contriving, and great dexterity in performing experiments. He had, indeed, very early applied himself to the prosecution of experimental science, and was one of the members of the small but distinguished body, who, during the civil wars, held private meetings for cultivating natural knowledge, on the plan of Bacon. They met first in London, as early as 1645, afterwards at Oxford, taking the name of the *Philosophic College*. Of them, when Charles the Second ascended the throne, was formed the Royal Society of London, incorporated by letters patent in 1662. No one was more useful than Boyle in communicating activity and vigour to the new institution. A real lover of knowledge, he was most zealous in the pursuit of it; and having thoroughly imbibed the spirit of Bacon, was an avowed enemy to the philosophy of Aristotle.

SECTION IV.

ASTRONOMY.

1. ANCIENT ASTRONOMY.

IT has already been remarked, that the ancients made more considerable advances in astronomy than in almost any other of the physical sciences. They applied themselves diligently to observe the heavens, and employed mathematical reasoning to connect together the insulated facts, which are the only objects of direct observation. The astronomer discovers nothing by help of his instruments, but that, at a given instant, a certain luminous point has a particular position in the heavens. The application of mathematics, and particularly of spherical trigonometry, enables him to trace out the precise tract of this luminous spot; to discover the rate of its motion, whether varied or uniform, and thus to resolve the first great problem which the science of astronomy involves, viz. to express the positions of the heavenly bodies, relatively to a given plane in functions of the time. The problem thus generally enunciated, comprehends all that is usually called by the name of descriptive or mathematical astronomy.

The explanation of the celestial motions, which naturally occurred to those who began the study of the heavens, was, that the stars are so many luminous points fixed in the surface of a sphere, having the earth in its centre, and revolving on an axis passing through that centre in the space of twenty-four hours. When it was observed that all the stars did not partake of this diurnal motion in the same degree, but that some were carried slowly towards the east, and that their paths estimated in that direction, after certain intervals of time, returned into themselves, it was believed that they were fixed in the surfaces of spheres, which revolved westward, more slowly than the sphere of the fixed stars. These spheres must be transparent, or made of some crystalline substance, and hence the name of the crystalline spheres, by which they were distinguished. This system, though it grew more complicated in proportion to the number and variety of the phenomena observed, was the system of Aristotle and Eudoxus, and, with a few exceptions, of all the philosophers of antiquity.

But when the business of observation came to be regularly pursued; when Timocharis and Aristillus, and their successors in the Alexandrian school, began to study the phenomena of the heavens, little was said of these orbs; and astronomers seemed only desirous of ascertaining the laws or the general facts concerning the planetary motions.

To do this, however, without the introduction of hypothesis, was certainly difficult, and probably was then impossible. The simplest and most natural hypothesis was, that the planets moved eastward in circles, and at a uniform rate. But when it was found that, instead of moving uniformly to the eastward, every one of them was subject to great irregularity, the motion eastward becoming at certain periods slower, and at length vanishing altogether, so that the planet became stationary, and afterwards acquiring a motion in the contrary direction, proceeded for a time toward the west, it was far from obvious how all these appearances could be reconciled with the idea of a uniform circular motion.

The solution of this difficulty is ascribed to Apollonius Pergæus, one of the greatest geometers of antiquity. He conceived that, in the circumference of a circle, having the earth for its centre, there moved the centre of another circle, in the circumference of which the planet actually revolved. The first of these circles was called the *deferent*, and the second the *epicycle*, and the motion in the circumference of each was supposed uniform. Lastly, it was conceived that the motion of the centre of the epicycle in the circumference of the deferent, and of the planet in that of the epicycle, were in opposite directions, the first being towards the east, and the second towards the west. In this way, the alterations from progressive to retrograde, with the intermediate stationary points, were readily explained, and Apollonius carried his investigation so far as to determine the ratio between the radius of the deferent, and that of the epicycle, from knowing the stations and retrogradations of any particular planet.

An object, which was then considered as of great importance to astronomy, was thus accomplished, viz. the production of a variable motion, or one which was continually changing both its rate and its direction from two uniform circular motions, each of which preserved always the same quantity and the same direction.

It was not long before another application was made of the method of epicycles. Hipparchus, the greatest astronomer of antiquity, and one of the inventors in science most justly entitled to admiration, discovered the inequality of the sun's apparent motion round the earth. To explain or to express this irregularity, the same observer imagined an epicycle of a small radius with its centre moving uniformly in the circumference of a large circle, of which the earth was the centre, while the sun revolved in the circumference of the small circle with the same angular velocity as this last, but in a contrary direction.

As other irregularities in the motions of the moon and of the planets were observed, other epicycles were introduced, and Ptolemy, in his *Alma-*

gest, enumerated all which then appeared necessary, and assigned to them such dimensions as enabled them to express the phenomena with accuracy. It is not to be denied that the system of the heavens became in this way extremely complicated; though, when fairly examined, it will appear to be a work of great ingenuity and research. The ancients, indeed, may be regarded as very fortunate in the contrivance of epicycles, because, by means of them, every inequality which can exist in the angular motion of a planet may be at least nearly represented. This I call fortunate, because, at the time when Apollonius introduced the epicycle, he had no idea of the extent to which his contrivance would go, as he could have none of the conclusions which the author of the *Mécanique Céleste* was to deduce from the principle of gravitation.

The same contrivance had another great advantage; it subjected the motions of the sun, the moon, and the planets, very readily to a geometrical construction, or an arithmetical calculation, neither of them difficult. By this means the prediction of astronomical phenomena, the calculation of tables, and the comparison of those tables with observation, became matters of great facility, on which facility, in a great measure, the progress of the science depended. It was on these circumstances, much more than on the simplicity with which it amused or deceived the imagination, that the popularity of this theory was founded; the ascendant which it gained over the minds of astronomers, and the resistance which, in spite of facts and observations, it was so long able to make to the true system of the world.

It does not appear that the ancient astronomers ever considered the epicycles and deferents which they employed in their system as having a physical existence, or as serving to *explain* the causes of the celestial motions. They seem to have considered them merely as mathematical diagrams, serving to *express* or to represent those motions as geometrical expressions of certain general facts, which readily furnished the rules of astronomical calculation.

The language in which Ptolemy speaks of the epicycles is not a little curious, and very conformable to the notion, that he considered them as merely the means of expressing a general law. After laying down the hypothesis of certain epicycles, and their dimensions, it is usual with him to add, "these suppositions will *save* the phenomena." *Save* is the literal translation of the Greek word, which is always a part of the verb Σωζειν, or some of its compounds. Thus, in treating of certain phenomena in the moon's motion, he lays down two hypotheses, by either of which they may be expressed; and he concludes, "In this way the similitude of the ratios, and the proportionality of the times, will be *saved* (*διασώζοιντο*) on both suppositions." * It is plain, from these words, that the astronomer did not here consider himself as describing any thing which actually existed, but as explaining two artifices, by either of which, certain irregularities in the moon's motion may be represented, in consistence with the principle of uniform velocity. The hypothesis does not relate to the explanation, but merely to the expression of the fact; it is first assumed, and its merit is then judged of synthetically, by its power to *save*, to reconcile, or to repre-

---

* Mathematica Syntaxis, Lib. IV. p. 223 of the Paris edition.——Milton, the extent and accuracy of whose erudition can never be too much admired, had probably in view this phraseology of Ptolemy, when he wrote the following lines:---

——" He his fabric of the Heavens
Hath left to their disputes, perhaps to move
His laughter at their quaint opinions wide
Hereafter, when they come to model Heaven
And calculate the stars, how they will wield
The mighty frame, how build, unbuild, contrive
To *save appearances*, how gird the sphere
With centrick and eccentrick scribbled o'er,
Cycle and epicycle, orb in orb."

The obsolete verb to *salve* is employed by Bacon, and many other of the old English writers in the same sense with Σωζειν in the work of Ptolemy here referred to. "The schoolmen were like the astronomers, who, to *salve* phenomena, framed to their conceit eccentricks and epicycles; so they, to *salve* the practice of the church, had devised a great number of strange positions."---Bacon.

sent appearances. At a time when the mathematical sciences extended little beyond the elements, and when problems which could not be resolved by circles and straight lines, could hardly be resolved at all, such artifices as the preceding were of the greatest value. They were even more valuable than the truth itself would have been in such circumstances; and nothing is more certain than that the real elliptical orbits of the planets, and the uniform description of areas, would have been very unseasonable discoveries at the period we are now treating of. The hypotheses of epicycles, and of centres of uniform motion, were well accommodated to the state of science, and are instances of a false system which has materially contributed to the establishment of truth.

2. Copernicus and Tycho.

On the revival of learning in Europe, astronomy was the first of the sciences which was regenerated. Such, indeed, is the beauty and usefulness of this branch of knowledge, that, in the thickest darkness of the middle ages, the study of it was never entirely abandoned. In those times of ignorance, it also derived additional credit from the assistance which it seemed to give to an imaginary and illusive science. Astrology, which has exercised so durable and extensive a dominion over the human mind, is coëval with the first observations of astronomy. In the middle ages, remarkable for the mixture of a few fragments of knowledge and truth in a vast mass of ignorance and error, it was assiduously cultivated, and, in conjunction with alchemy and magic, shared the favour of the people, and the patronage of the great. During the thirteenth and fourteenth centuries, it was taught in the universities of Italy, and professors were appointed, at Padua and Bologna, to instruct their pupils in the influence of the stars. Everywhere through Europe the greatest respect was shown for this system of imposture, and they who saw the deceit most clearly, could not always avoid the disgrace of being the instruments of it. Astronomy, however, profited by the illusion, and was protected for the great assistance whish it seemed to afford to a science more importan than itself.

Of those who cultivated astronomy, many were infected by this weakness, though some were completely superior to it. Alphonso, the King of Castile, was among the latter. He flourished about the middle of the thirteenth century, and was remarkable for such freedom of thought, and such boldness of language, as it required his royal dignity to protect. He applied himself diligently to the study of astronomy ; he perceived the inaccuracy of Ptolemy's tables, and endeavoured to correct their errors by new tables of his own. These, in the course of the next age, were found to have receded from the heavens, and it became more and more evident that astronomers had not yet discovered the secret of the celestial motions.

Two of the men who, in the fifteenth century, contributed the most to the advancement of astronomical science, Purbach and Regiomontanus, were distinguished also for their general knowledge of the mathematics. Purbach was fixed at Vienna by the patronage of the Emperor Frederick the Third, and devoted himself to astronomical observation. He published a new edition of the *Almagest*, and, though he neither understood Greek nor Arabic, his knowledge of the subject enabled him to make it much more perfect than any of the former translations. He is said to have been the first who applied the plummet to astronomical instruments; but this must not be understood strictly, for some of Ptolemy's instruments, the parallactic for instance, were placed perpendicularly by the plumb line.

Regiomontanus was the disciple of Purbach, and is still more celebrated than his master. He was a man of great learning and genius, most ardent for the advancement of knowledge, and particularly devoted to astronomy. To him we owe the introduction of decimal fractions, which completed our arithmetical notation, and formed the second of the three steps by which, in modern times, the science of numbers has been so greatly improved.

In the list of distinguished astronomers, the name of Copernicus comes next, and stands at the head of those men, who, bursting the fetters of prejudice and authority, have established truth on the basis of experience and observation. He was born at Thorn in Prussia, in 1473; he studied at the university of Cracow, being intended at first for a physician, though he afterwards entered into the church. A decided taste for astronomy led him early to the study of the science in which he was destined to make such an entire revolution, and as soon as he found himself fixed and independent, he became a diligent and careful observer.

It would be in the highest degree interesting to know by what steps he was led to conceive the bold system which removes the earth from the centre of the world, and ascribes to it a twofold motion. It is probable that the complication of so many epicycles and deferents as were necessary, merely to express the laws of the planetary motions, had induced him to think of all the possible suppositions which could be employed for the same purpose, in order to discover which of them was the simplest.

It appears extraordinary, that so natural a thought should have occurred, at so late a period, for the first, or nearly for the first time. We are assured, by Copernicus himself, that one of the first considerations which offered itself to his mind, was the effect produced by the motion of a spectator, in transferring that motion to the objects observed, but ascribing to it an opposite direction.* From this principle it immediately followed, that the rotation of the earth on an axis, from west to east, would produce the apparent motion of the heavens in the direction from east to west.

In considering some of the objections which might be made to the system of the earth's motion, Copernicus reasons with great soundness, though he is not aware of the full force of his own argument. Ptolemy had alleged, that, if the earth were to revolve on its axis, the violence of the motion would be sufficient to tear it in pieces, and to dissipate the parts. This argument, it is evident, proceeds on a confused notion of a centrifugal force, the effect of which the Egyptian astronomer overrated, as much as he undervalued the firmness and solidity of the earth. Why, says Copernicus, was he not more alarmed for the safety of the heavens, if the diurnal revolution be ascribed to them, as their motion must be more rapid, in pro-

* Astronomia Instaurata, Lib. I. cap. 5.

portion as their magnitude is greater? The argument here suggested, now that we know how to measure centrifugal force, and to compare it with others, carries demonstrative evidence with it, because that force, if the diurnal revolution were really performed by the heavens, would be such, as the forces which hold together the frame of the material world would be wholly unable to resist.

There are, however, in the reasonings of Copernicus, some unsound parts, which show, that the power of his genius was not able to dispel all the clouds which in that age hung over the human mind, and that the unfounded distinctions of the Aristotelian physics sometimes afforded arguments equally fallacious to him and to his adversaries. One of his most remarkable physical mistakes was his misconception with respect to the parallelism of the earth's axis; to account for which, he thought it necessary to assume, in addition to the earth's rotation on an axis, and revolution round the sun, the existence of a third motion altogether distinct from either of the others. In this he was mistaken; the axis naturally retains its parallelism, and it would require the action of a force to make it do otherwise. This, as Kepler afterwards remarked, is a consequence of the *inertia* of matter; and for that reason, he very justly accused Copernicus of not being fully acquainted with his own riches.

The first edition of the Astronomia Instaurata,

the publication of which was solicited by Cardinal Schoenberg, and the book itself dedicated to the Pope, appeared in 1543, a few days before the death of the author. * Throughout the whole book, the new doctrine was advanced with great caution, as if from a presentiment of the opposition and injustice which it was one day to experience. At first, however, the system attracted little notice, and was rejected by the greater part even of astronomers. It lay fermenting in secret with other new discoveries for more than fifty years, till, by the exertions of Galileo, it was kindled into so bright a flame as to consume the philosophy of Aristotle, to alarm the hierarchy of Rome, and to threaten the existence of every opinion not founded on experience and observation.

After Copernicus, Tycho Brahé was the most distinguished astronomer of the sixteenth century. An eclipse of the sun which he witnessed in 1560, when he was yet a very young man, by the exactness with which it answered to the prediction, impressed him with the greatest reverence for a science which could see so far and so distinctly into the future, and from that moment he was seized with the strongest desire of becoming acquainted

* The first edition here alluded to bore the title of *De Revolutionibus Orbium Celestium, Libri VI.* See Biographie Universelle, Art. Copernicus.—E.

with it. Here, indeed, was called into action a propensity nearly allied both to the strength and the weakness of the mind of this extraordinary man, the same that attached him, on one hand, to the calculations of astronomy, and, on the other, to the predictions of judicial astrology.

In yielding himself up, however, to his love of astronomy, he found that he had several difficulties to overcome. He belonged to a class in society elevated, in the opinion of that age, above the pursuit of knowledge, and jealous of the privilege of remaining ignorant with impunity. Tycho was of a noble family in Denmark, so that it required all the enthusiasm and firmness inspired by the love of knowledge, to set him above the prejudices of hereditary rank, and the opposition of his relations. He succeeded, however, in these objects, and also in obtaining the patronage of the King of Denmark, by which he was enabled to erect an observatory, and form an establishment in the island of Huena, such as had never yet been dedicated to astronomy. The instruments were of far greater size, more skilfully contrived, and more nicely divided, than any that had yet been directed to the heavens. By means of them, Tycho could measure angles to ten seconds, which may be accounted sixty times the accuracy of the instruments of Ptolemy, or of any that had belonged to the school of Alexandria.

Among the improvements which he made in the art of astronomical observation, was that of verifying the instruments, or determining their errors by actual observation, instead of trusting, as had been hitherto done, to the supposed infallibility of the original construction.

One of the first objects to which the Danish astronomer applied himself was the formation of a new catalogue of the fixed stars. That which was begun by Hipparchus, and continued by Ptolemy, did not give the places of the stars with an accuracy nearly equal to that which the new instruments were capable of reaching ; and it was besides desirable to know whether the lapse of twelve centuries had produced any unforeseen changes in the heavens.

The great difficulty in the execution of this work arose from the want of a direct and easy method of ascertaining the distance of one heavenly body due east or west of another. The distance north or south, either from one another or from a fixed plane, that of the equator, was easily determined by the common method of meridian altitudes, the equator being a plane which, for any given place on the earth's surface, retains always the same position. But no plane extending from north to south, or passing through the poles, retains a fixed position with respect to an observer, and, therefore, the same way of measuring distances from such a plane cannot be applied. The natural substitute is the measure of time; the interval between the passage of two stars over the meridian, bearing the same proportion to twenty-four hours, that the arch which measures their distance perpendicular to the meridian, or their difference of right ascension, does to four right angles.

An accurate measure of time, therefore, would answer the purpose, but such a measure no more existed in the age of Tycho, than it had done in the days of Hipparchus or Ptolemy. These ancient astronomers determined the longitude of the fixed stars by referring their places to those of the moon, the longitude of which, for a given time, was known from the theory of her motions. Thus they were forced to depend on the most irregular of all the bodies in the heavens, for ascertaining the positions of the most fixed, those which ought to have been the basis of the former, and of so many other determinations. Tycho made use of the planet Venus instead of the moon, and his method, though more tedious, was more accurate than that of the Greek astronomers. His catalogue contained the places of 777 fixed stars.

The irregularities of the moon's motions were his next subjects of inquiry. The ancients had discovered the inequality of that planet depending on the eccentricity of the orbit, the same which is now called the equation of the centre.* Ptolemy had added the knowledge of another inequality in the moon's motion, to which the name of the evection has been given, amounting to an increase of the former equation at the quarters, and a diminution of it at the times of new and full moon. Tycho discovered another inequality, which is greatest at the octants, and depends on the difference between the longitude of the moon and that of the sun. A fourth irregularity to which the moon's motion is subject, depending wholly on the sun's place, was known to Tycho, but included among the sun's equations. Besides, these observations made him acquainted with the changes in the inclination of the plane of the moon's orbit; and, lastly, with the irregular motion of the nodes, which, instead of being always retrograde at the same rate, are subject to change that rate, and even to become progressive according to their situation in respect of the sun. These are the only inequalities of the moon's motion known before the theory of Gravitation, and, except the two first, are all the discoveries of Tycho.

The atmospherical refraction, by which the heavenly bodies are made to appear more elevated

* The allowance made for any such inequality, when the place of a planet is to be computed for a given time, is called an *equation* in the language of astronomy.

above the horizon than they really are, was suspected before the time of this astronomer, but not known with certainty to exist. He first became acquainted with it by finding that the latitude of his observatory, as determined from observations at the solstices, and from observations of the greatest and least altitudes of the circumpolar stars, always differed about four minutes. The effect of refraction he supposed to be 34′ at the horizon, and to diminish from thence upward, till at 45° it ceased altogether. This last supposition is erroneous, but at 45° the refraction is less than 1′, and probably was not sensible in the altitudes measured with his instruments, or not distinguishable from the errors of observation. An instrument which he contrived on purpose to make the refraction distinctly visible, shows the scale on which his observatory was furnished. It was an equatorial circle of ten feet diameter, turning on an axis parallel to that of the earth. With the sights of this equatorial he followed the sun on the day of the summer solstice, and found, that, as it descended towards the horizon, it rose above the plane of the instrument. At its setting, the sun was raised above the horizon by more than its own diameter.

The comet of 1570 was carefully observed by Tycho, and gave rise to a new theory of those bodies. He found the horizontal parallax to be 20′, so that the comet was nearly three times as far off as the moon. He considered comets, therefore, as bodies placed far beyond the reach of our atmosphere, and moving round the sun. This was a severe blow to the physics of Aristotle, which regarded comets as meteors generated in the atmosphere. His observation of the new star in 1572, was no less hostile to the argument of the same philosopher, which maintained, that the heavens are a region in which there is neither generation nor corruption, and in which existence has neither a beginning nor an end.

Yet Tycho, with this knowledge of astronomy, and after having made observations more numerous and accurate than all the astronomers who went before him, continued to reject the system of Copernicus, and to deny the motion of the earth. He was, however, convinced that the earth is not the centre about which the planets revolve, for he had himself observed Mars, when in opposition, to be nearer to the earth than the earth was to the sun, so that, if the planets were ranged as in the Ptolemaic system, the orbit of Mars must have been within the orbit of the sun. He therefore imagined the system still known by his name, according to which the sun moves round the earth, and is at the same time the centre of the planetary motions. It cannot be denied, that the phenomena purely astronomical may be accounted for on this hypothesis, and that the objections to it are rather derived from physical and mechanical considerations than from the appearances themselves. It is simpler than the Ptolemaic system, and free from its inconsistencies; but it is more complex than the Copernican, and, in no respect, affords a better explanation of the phenomena. The true place of the Tychonic system is between the two former; an advance beyond the one, and a step short of the other; and such, if the progress of discovery were always perfectly regular, is the place which it would have occupied in the history of the science. If Tycho had lived before Copernicus, his system would have been a step in the advancement of knowledge; coming after him, it was a step backward.

It is not to his credit as a philosopher to have made this retrograde movement, yet he is not altogether without apology. The physical arguments in favour of the Copernican system, founded on the incongruity of supposing the greater body to move round the smaller, might not be supposed to have much weight, in an age when the equality of action and reaction was unknown, and when it was not clearly understood that the sun and the planets act at all on one another. The arguments, which seem, in the judgment of Tycho, to have balanced the simplicity of the Copernican system, were founded on certain texts of Scripture, and on the difficulty of reconciling the motion of the earth with the sensations which we experience at its surface, or the phenomena which we observe, the same, in all respects, as if the earth were at rest. The experiments and reasonings of Galileo had not yet instructed men in the *inertia* of matter, or in the composition of motion; and the followers of Copernicus reasoned on principles which they held in common with their adversaries. A ball, it was said by the latter, dropt from the mast-head of a ship under sail, does not fall at the foot of the mast, but somewhat behind it; and, in the same manner, a stone dropt from a high tower would not fall, on the supposition of the earth's motion, at the bottom of the tower, but to the west of it, the earth, during its fall, having gone eastward from under it. The followers of Copernicus were not yet provided with the true answer to this objection, viz. that the ball does actually fall at the bottom of the mast. It was admitted that it must fall behind it, because the ball was no part of the ship, and that the motion forward was not natural, either to the ship or to the ball. The stone, on the other hand, let fall from the top of the tower, was a part of the earth; and, therefore, the diurnal and annual revolutions which were natural to the earth, were also natural to the stone; the stone would, therefore, retain the same motion with the tower, and strike the ground precisely at the bottom of it.

It must be confessed, that neither of these logicians had yet thoroughly awakened from the dreams of the Aristotelian metaphysics, but men were now in possession of the truth, which was finally to break the spell, and set the mind free from the fetters of prejudice and authority. Another charge, against which it is more difficult to defend Tycho, is his belief in the predictions of astrology. He even wrote a treatise in defence of this imaginary art, and regulated his conduct continually by its precepts. Credulity, so unworthy of a man deeply versed in real science, is certainly to be set down less to his own account than to that of the age in which he lived.

3. Kepler and Galileo.

Kepler followed Tycho, and in his hands astronomy underwent a change only second to that which it had undergone in the hands of Copernicus. He was born in 1571. He early applied himself to study and observe the heavens, and was soon distinguished as an inventor. He began with taking a more accurate view of astronomical refraction than had yet been done, and he appears to have been the first who conceived that there must be a certain fixed law which determined the quantity of it, corresponding to every altitude, from the horizon to the zenith. The application of the principles of optics to astronomy, and the accurate distinction between the optical and real inequalities of the planets, are the work of the same astronomer. It was by the views thus presented that he was led to the method of constructing and calculating eclipses, by means of projections, without taking into consideration the diurnal parallax. These are valuable improvements, but they were, however, obscured by the greatness of his future discoveries.

The planes of the orbits of the planets were naturally, in the Ptolemaic system, supposed to pass through the earth, and the reformation of Copernicus did not go so far as to change the notions on that subject which had generally been adopted. Kepler observed that the orbits of the planets are in planes passing through the sun, and that, of consequence, the lines of their nodes all intersect in the centre of that luminary. This discovery contributed essentially to those which followed.

The oppositions of the planets, or their places when they pass the meridian at midnight, offer the most favourable opportunities for observing them, both because they are at that time nearest to the earth, and because their places seen from thence are the same as if they were seen from the sun. The true time of the opposition had, however, been till now mistaken by astronomers, who held it to be at the moment when the apparent place of the planet was opposite to the mean place of the sun. It ought, however, to have been, when the apparent places of both were opposed to one another. This reformation was proposed by Kepler, and, though strenuously resisted by Tycho, was finally received.

Having undertaken to examine the orbit of Mars, in which the irregularities are most considerable, Kepler discovered, by comparing together seven oppositions of that planet, that its orbit is elliptical; that the sun is placed in one of the foci; and that there is no point round which the angular motion is uniform. In the pursuit of this inquiry he found that the same thing is true of the earth's orbit round the sun; hence by analogy it was reasonable to think, that all the planetary orbits are elliptical, having the sun in their common focus.

The industry and patience of Kepler, in this investigation, were not less remarkable than his ingenuity and invention. Logarithms were not yet known, so that arithmetical computation, when pushed to great accuracy, was carried on at a vast expence of time and labour. In the calculation of every opposition of Mars, the work filled ten folio pages, and Kepler repeated each calculation ten times, so that the whole work for each opposition extended to one hundred such pages; seven oppositions thus calculated produced a large folio volume.

In these calculations the introduction of hypotheses was unavoidable, and Kepler's candour in rejecting them, whenever they appeared erroneous, without any other regret than for the time which they had cost him, cannot be sufficiently admired. He began with hypothesis, and ended with rejecting every thing hypothetical. In this great astronomer we find genius, industry, and candour, all uniting together as instruments of investigation.

Though the angular motion of the planet was not found to be uniform, it was discovered that a very simple law connected that motion with the rectilineal distance from the sun, the former being every where inversely as the square of the latter; and hence it was easy to prove, that the area described by the line drawn from the planet to the sun increased at a uniform rate, and, therefore, that any two such areas were proportional to the times in which they were described. The picture presented of the heavens was thus, for the first time, cleared of every thing hypothetical.

The same astronomer was perhaps the first person who conceived that there must be always a law capable of being expressed by arithmetic or geometry, which connects such phenomena as have a physical dependence on one another. His conviction of this truth, and the delight which he appears to have experienced in the contemplation of such laws, led him to seek, with great eagerness, for the relation between the periodical times of the planets, and their distances from the sun. He seems, indeed, to have looked towards this object with such earnestness, that, while it was not attained, he regarded all his other discoveries as incomplete. He at last found, infinitely to his satisfaction, that in any two planets, the squares of the times of the revolution are as the cubes of their mean distances from the sun. This beautiful and simple law had a value beyond what Kepler could possibly conceive; yet a sort of scientific instinct instructed him in its great importance. He has marked the year and the day when it became known to him; it was on the 8th of May 1618; and perhaps philosophers will agree that there are few days in the scientific history of the world which deserve so well to be remembered.

These great discoveries, however, were not much attended to by the astronomers of that period, or by those who immediately followed. They were but little considered by Gassendi,—they were undervalued by Riccioli,—and were never mentioned by Descartes. It was an honour reserved for Newton to estimate them at their true value.

Indeed, the discoveries of Kepler were at first so far from being duly appreciated, that they were objected to, not for being false, but for offering to astronomers, in the calculation of the place of a planet in its orbit, a problem too difficult to be resolved by elementary geometry. To cut the area of a semi-ellipsis in a given ratio by a line drawn through the focus, is the geometrical problem into which he showed that the above inquiry ultimately resolved. As if he had been answerable for the proceedings of nature, the difficulty of this question was considered as an argument against his theory, and he himself seems somewhat to have felt it as an objection, especially when he found that the best solution he could obtain was no more than an approximation.. With all his power of invention, Kepler was a mathematician inferior to many of that period; and though he displayed great ability in the management of this difficult investigation, his solution fell very far short of the simplicity which it was afterwards found capable of attaining.

In addition to all this, he rendered another very important service to the science of astronomy and to the system of Copernicus. Copernicus, it has been already mentioned, had supposed that a force was necessary to enable the earth to preserve the parallelism of its axis during its revolution round the sun. He imagined, therefore, that a third motion belonged to the earth, and that, besides turning on its axis and revolving round the sun, it had another movement by which its axis was preserved always equally inclined to the ecliptic. Kepler was the first to observe that this third motion was quite superfluous, and that the parallelism of the earth's axis, in order to be preserved, required nothing but the absence of all force, as it necessarily proceeded from the inertia of matter, and its tendency to persevere in a state of uniform motion. Kepler had a clear idea of the inertia of body ; he was the first who employed the term ; and, considering all motion as naturally rectilineal, he concluded that when a body moves in a curve, it is drawn or forced out of the straight line by the action of some cause, not residing in itself. Thus he prepared the way for physical astronomy, and in these ideas he was earlier than Descartes.

The discoveries of Kepler were secrets extorted from nature by the most profound and laborious research. The astronomical discoveries of Galileo, more brilliant and imposing, were made at a far less expence of intellectual labour. By this it is not meant to say that Galileo did not possess, and did not exert intellectual powers of the very highest order, but it was less in his astronomical discoveries that he had occasion to exert them, than in those which concerned the theory of motion. The telescope turned to the heavens for the first time, in the hands of a man far inferior to the Italian philosopher, must have unfolded a series of wonders to astonish and delight the world.

It was in the year 1609 that the news of a discovery, made in Holland, reached Galileo, viz. that two glasses had been so combined, as greatly to magnify the objects seen through them. More was not told, and more was not necessary to awaken a mind abundantly alive to all that interested the progress either of science or of art. Galileo applied himself to try various combinations of lenses, and he quickly fell on one which made objects appear greater than when seen by the naked eye, in the proportion of three to one. He soon improved on this construction, and found one which magnified thirty-two times, nearly as much as the kind of telescope he used is capable of. That telescope was formed of two lenses; the lens next the object convex, the other concave; the objects were presented upright, and magnified in their lineal dimensions in the proportion just assigned.

Having tried the effect of this combination on terrestrial objects, he next directed it to the moon. What the telescope discovers on the ever-varying face of that luminary, is now well-known, and needs not to be described; but the sensations which the view must have communicated to the philosopher who first beheld it, may be conceived more easily than expressed. To the immediate impression which they made upon the sense, to the wonder they excited in all who saw them, was added the proof, which, on reflection, they afforded, of the close resemblance between the earth and the celestial bodies, whose divine nature had been so long and so erroneously contrasted with the ponderous and opaque substance of our globe. The earth and the planets were now proved to be bodies of the same kind, and views were entertained of the universe, more suitable to the simplicity and the magnificence of nature.

When the same philosopher directed his telescope to the fixed stars, if he was disappointed at finding their magnitudes not increased, he was astonished and delighted to find them multiplied in so great a degree, and such numbers brought into view, which were invisible to the naked eye. In Jupiter he perceived a large disk, approaching in size to the moon. Near it, as he saw it for the first time, were three luminous points ranged in a straight line, two of them on one side of the planet, and one on the other. This occasioned no surprise, for they might be small stars not visible to the naked eye, such as he had already discovered in great numbers. By observing them, however, night after night, he found these small stars to be four in number, and to be moons or satellites, accompanying Jupiter, and revolving round him, as the moon revolves round the earth.

The eclipses of these satellites, their conjunctions with the planet, their disappearance behind his disk, their periodical revolutions, and the very problem of distinguishing them from one another, offered, to an astronomer, a series of new and interesting observations.

In Saturn he saw one large disk, with two smaller ones very near it, and diametrically opposite, and always seen in the same places; but more powerful telescopes were required before these appearances could be interpreted.

The horned figure of Venus, and the gibbosity of Mars, added to the evidence of the Copernican system, and verified the conjectures of its author, who had ventured to say, that, if the sense of sight were sufficiently powerful, we should see Mercury and Venus exhibiting phases similar to those of the moon.

The spots of the sun derived an interest from their contrast with the luminous disk over which they seemed to pass. They were found to have such regular periods of return, as could be derived only from the motion of the disk itself; and thus the sun's revolution on his axis, and the time of that revolution, were clearly ascertained.

This succession of noble discoveries, the most splendid, probably, which it ever fell to the lot of one individual to make, in a better age would have entitled its author to the admiration and gratitude of the whole scientific world, but was now viewed from several quarters with suspicion and jealousy. The ability and success with which Galileo had laboured to overturn the doctrines of Aristotle and the schoolmen, as well as to establish the motion of the earth, and the immobility of the sun, had excited many enemies. There are always great numbers who, from habit, indolence, or fear, are the determined supporters of what is established, whether in practice or in opinion. To these the constitution of the universities of Europe, so entirely subjected to the church, had added a numerous and learned phalanx, interested to preserve the old systems, and to resist all innovations which could endanger their authority or their repose. The church itself was roused to action, by reflecting that it had staked the infallibility of its judgments on the truth of the very opinions which were now in danger of being overthrown. Thus was formed a vast combination of men, not very scrupulous about the means which they used to annoy their adversaries; the power was entirely in their hands, and there was nothing but truth and reason to be opposed to it.

The system of Copernicus, however, while it remained obscure, and known only to astronomers, created no alarm in the church. It had even been ushered into the world at the solicitation of a cardinal, and under the patronage of the Pope; but when it became more popular, when the ability and acuteness of Galileo were enlisted on its side, the consequences became alarming; and it was determined to silence by force an adversary who could not be put down by argument. His dialogues contained a full exposition of the evidence of the earth's motion, and set forth the errors of the old, as well as the discoveries of the new philosophy, with great force of reasoning, and with the charms of the most lively eloquence. They are written, indeed, with such singular felicity, that one reads them at the present day, when the truths contained in them are known and admitted, with all the delight of novelty, and feels one's self carried back to the period when the telescope was first directed to the heavens, and when the earth's motion, with all its train of consequences, was proved for the first time. The author of such a work could not be forgiven. Galileo, accordingly, was twice brought before the Inquisition. The first time a council of seven cardinals pronounced a sentence which, for the sake of those disposed to believe that power can subdue truth, ought never to be forgotten: "That to maintain the sun to be immoveable, and without local motion, in the centre of the world, is an absurd proposition, false in philosophy, heretical in religion, and contrary to the testimony of Scripture. That it is equally absurd and false in philosophy to assert that the earth is not immoveable in the centre of the world, and, considered theologically, equally erroneous and heretical."

These seven theologians might think themselves officially entitled to decide on what was heretical or orthodox in faith, but that they should determine what was true or false in philosophy, was an insolent invasion of a territory into which they had no right to enter, and is a proof how ready men are to suppose themselves wise, merely because they happen to be powerful. At this time a promise was extorted from Galileo, that he would not teach the doctrine of the earth's motion, either by speaking or by writing. To this promise he did not conform. His third dialogue, published, though not till long afterwards, contained such a full display of the beauty and simplicity of the new system, and such an exposure of the inconsistencies of Ptolemy and Tycho, as completed the triumph of Copernicus.

In the year 1663, Galileo, now seventy years old, being brought before the Inquisition, was forced solemnly to disavow his belief in the earth's motion; and condemned to perpetual imprisonment, though the sentence was afterwards mitigated, and he was allowed to return to Florence.* The Court of Rome was very careful to publish this second recantation all over Europe, thinking, no doubt, that

* He was thrown into prison previously to his trial, and attempts were made to render him obnoxious to the people. From the text of a priest who preached against him, we may judge of the wit and the sense with which this persecution was conducted. *Viri Galilæi quid statis in cœlum suspicientes?*

it was administering a complete antidote to the belief of the Copernican system. The sentence, indeed, appears to have pressed very heavily on Galileo's mind, and he never afterwards either talked or wrote on the subject of astronomy. Such was the triumph of his enemies, on whom ample vengeance would have long ago been executed, if the indignation and contempt of posterity could reach the mansions of the dead.

Conduct like this, in men professing to be the ministers of religion and the guardians of truth, can give rise to none but the most painful reflections. That an aged philosopher should be forced, laying his hand on the sacred Scriptures, to disavow opinions which he could not cease to hold without ceasing to think, was as much a profanation of religion, as a violation of truth and justice. Was it the act of hypocrites, who considered religion as a state engine, or of bigots, long trained in the art of believing without evidence, or even in opposition to it? These questions it were unnecessary to resolve; but one conclusion cannot be denied, that the indiscreet defenders of religion have often proved its worst enemies.

At length, however, by the improvements, the discoveries, and the reasonings, first of Kepler, and then of Galileo, the evidence of the Copernican system was fully developed, and nothing was wanting to its complete establishment, but time sufficient to allow opinion to come gradually round, and to give men an opportunity of studying the arguments placed before them. Of the adherents of the old system, many had been too long habituated to it to change their views; but as they disappeared from the scene, they were replaced by young astronomers, not under the influence of the same prejudices, and eager to follow doctrines which seemed to offer so many new subjects of investigation. In the next generation the systems of Ptolemy and Tycho had no followers.

It was not astronomy alone which was benefited by this revolution, and the discussions to which it had given rise. A new light, as already remarked, was thrown on the physical world, and the curtain was drawn aside which had so long concealed the great experiment, by which nature herself manifests, at every instant, the inertia of body, and the composition of forces. To reconcile the real motion of the earth with its appearance of rest, and with our feeling of its immobility, required such an examination of the nature of motion, as discovered, if not its essence, at least its most general and fundamental properties. The whole science of rational mechanics profited, therefore, essentially by the discovery of the earth's motion.

A great barrier to philosophic improvement had arisen from the separation so early made, and so strenuously supported in the ancient systems, between terrestrial and celestial substances, and between the laws which regulate motion on the earth, and in the heavens. This barrier was now entirely removed; the earth was elevated to the rank of a planet; the planets were reduced to the condition of earths, and by this mutual approach, the same rules of interpretation became applicable to the phenomena of both. Principles derived from experiments on the earth, became guides for the analysis of the heavens, and men were now in a situation to undertake investigations, which the most hardy adventurer in science could not before have dared to imagine. Philosophers had ascended to the knowledge of the affinities which pervade all nature, and which mark so strongly both the wisdom and unity of its author.

The light thus struck out darted its rays into regions the most remote from physical inquiry. When men saw opinions entirely disproved, which were sanctioned by all antiquity, and by the authority of the greatest names, they began to have different notions of the rules of evidence, of the principles of philosophic inquiry, and of the nature of the mind itself. It appeared that science was destined to be continually progressive; provided it was taken for an inviolable maxim, that all opinion must be ultimately amenable to experience and observation.

It was no slight addition to all these advantages,

that, in consequence of the discussions from which Galileo had unhappily been so great a sufferer, the line was at length definitely drawn which was to separate the provinces of faith and philosophy from one another. It became a principle, recognised on all hands, that revelation, not being intended to inform men of those things which the unassisted powers of their own understanding would in time be able to discover, had, in speaking of such matters, employed the language and adopted the opinions of the times; and thus the magic circle by which the priest had endeavoured to circumscribe the inquiries of the philosopher entirely disappeared. The reformation in religion which was taking place about the same time, and giving such energy to the human mind, contributed to render this emancipation more complete, and to reduce the exorbitant pretensions of the Romish church. The prohibition against believing in the true system of the world either ceased altogether, or was reduced to an empty form, by which the affectation of infallibility still preserves the memory of its errors. *

* The learned fathers who have, with so much ability, commented on the Principia of Newton, have prefixed to the third book this remarkable declaration:—" Newtonus in hoc tertio libro telluris motæ hypothesin assumit. Auctoris propositiones aliter explicari, non poterant nisi eâdam factâ hypothesi. Hinc alienam coacti sumus gerere personam. *Ceterum latis a summis Pontificibus contra telluris motum Decre-*

4. Descartes, Huygens, &c.

Descartes flourished about this period, and has the merit of being the first who undertook to give an explanation of the celestial motions, or who formed the great and philosophic conception of reducing all the phenomena of the universe to the same law. The time was now arrived when, from the acknowledged assimilation of the planets to the earth, this might be undertaken with some reasonable prospect of success. No such attempt had hitherto been made, unless the crystalline spheres or homocentric orbs of the ancients are to be considered in that light. The conjectures of Kepler about a kind of animation, and of organic structure, which pervaded the planetary regions, were too vague and indefinite, and too little analogous to any thing known on the earth, to be entitled to the name of a theory. To Descartes, therefore, belongs the honour of being the first who ventured on the solution of the most arduous problem which the material world offers to the consideration of philosophy. For this solution he sought no other data than *matter* and *motion*, and with them alone

tis nos obsequi profitemur." There is an archness in the last sentence, that looks as if the authors wanted to convey meanings that would differ according to the orthodoxy of the readers.

proposed to explain the structure and constitution of the universe. The matter which he required, too, was of the simplest kind, possessing no properties but extension, impenetrability, and inertia. It was matter in the abstract, without any of its peculiar or distinguishing characters. To explain these characters, was indeed a part of the task which he proposed to himself, and thus, by the simplicity of his assumptions, he added infinitely to the difficulty of the problem which he undertook to resolve.

The matter thus constituted was supposed to fill all space, and its parts, both great and small, to be endued with motion in an infinite variety of directions. From the combination of these, the rectilineal motion of the parts become impossible; the atoms or particles of matter were continually diverted from the lines in which they had begun to move; so that circular motion and centrifugal force originated from their action on one another. Thus matter came to be formed into a multitude of vortices, differing in extent, in velocity, and in density; the more subtile parts constituting the real vortex, in which the denser bodies float, and by which they are pressed, though not equally, on all sides.

Thus the universe consists of a multitude of vortices, which limit and circumscribe one another. The earth and the planets are bodies carried round in the great vortex of the solar system; and by the pressure of the subtile matter, which circulates with great rapidity, and great centrifugal force, the denser bodies, which have less rapidity, and less centrifugal force, are forced down toward the sun, the centre of the vortex. In like manner, each planet is itself the centre of a smaller vortex, by the subtile matter of which the phenomena of gravity are produced, just as with us at the surface of the earth.

The gradation of smaller vortices may be continued in the same manner, to explain the cohesion of the grosser bodies, and their other sensible qualities. .But I forbear to enter into the detail of a system, which is now entirely exploded, and so inconsistent with the views of nature which have become familiar to every one, that such details can hardly be listened to with patience. Indeed, the theory of vortices did not explain a single phenomenon in a satisfactory manner, nor is there a truth of any kind which has been brought to light by means of it. None of the peculiar properties of the planetary orbits were taken into the account; none of the laws of Kepler were considered; nor was any explanation given of those laws, more than of any other that might be imagined. The philosophy of Descartes could explain all things equally well, and might have been accommodated to the systems of Ptolemy or Tycho, just as well as to that of Copernicus. It forms, therefore, no link in the chain of physical discovery; it served the cause of truth only by exploding errors more pernicious than its own; by exhausting a source of deception, which might have misled other adventurers in science, and by leaving a striking proof how little advancement can be made in philosophy, by pursuing any path but that of experiment and induction. Descartes was, nevertheless, a man of great genius, a deep thinker, of enlarged views, and entirely superior to prejudice. Yet, in as far as the explanation of astronomical phenomena is concerned, (and it was his main object,) he did good only by showing in what quarter the attempt could not be made with success; he was the forlorn hope of the new philosophy, and must be sacrificed for the benefit of those who were to follow.

Gassendi, the contemporary and countryman of Descartes, possessed great learning, with a very clear and sound understanding. He was a good observer, and an enlightened advocate of the Copernican system. He explained, in a very satisfactory manner, the connection between the laws of motion and the motion of the earth, and made experiments to show, that a body carried along by another acquires a motion which remains after it has ceased to be so carried. Gassendi first observed the transit of a planet over the disk of the sun, —that of Mercury, in 1631. Kepler had predicted this transit, but did not live to enjoy a spectacle which afforded so satisfactory a proof of the truth of his system, and of the accuracy of his astronomical tables.

The first transit of Venus, which was observed, happened a few years later, in 1639, when it was seen in England by Horrox, and his friend Crabtree, and by them only. Horrox, who was a young man of great genius, had himself calculated the transit, and foretold the time very accurately, though the astronomical tables of that day gave different results, and those of Kepler, in which he confided the most, were, in this instance, considerably in error. Horrox has also the merit of being among the first who rightly appreciated the discoveries of the astronomer just named. He had devoted much time to astronomical observation, and, though he died very young, he left behind him some preparations for computing tables of the moon, on a principle which was new, and which Newton himself thought worthy of being adopted in his theory of the inequalities of that planet.

The first complete system of astronomy, in which the elliptic orbits were introduced, was the *Astronomia Philolaica* of Bullialdus, (Bouillaud,) published in 1645. They were introduced, however, with such hypothetical additions, as show that the idea of a centre of uniform motion had not yet entirely disappeared. It is an idea, indeed, which gives considerable relief to the imagination, and it besides leads to methods of calculation more simple than the true theory, and Bullialdus may have flattered himself that they were sufficiently exact. He conceives the elliptic orbit as a section of an oblique cone, the axis of which passes through the superior focus of the ellipse, while the planet moves in its circumference in such a manner, that a plane passing through it and through the axis, shall be carried round with a uniform angular velocity. It is plain that the cone and its axis are mere fictions, arbitrarily assumed, and not even possessing the advantage of simplicity. The author himself departs from this hypothesis, and calculates the places of a planet, on the supposition that it moves in the circumference of an epicycle, and the epicycle in the circumference of an eccentric deferent, both angular motions being uniform, that of the planet in the epicycle being retrograde, and double the other. The figure thus described may be shown to be an ellipse, but the line drawn from the planet to the focus does not cut off areas proportional to the time.

An hypothesis advanced by Ward, Bishop of Salisbury, was simpler and more accurate than that of the French astronomer. According to it, the line drawn from a planet to the superior focus of its elliptic orbit, turns with a uniform angular velocity round that point. In orbits of small eccentricity, this is nearly true, and almost coincides in such cases with Kepler's principle of the uniform description of areas. Dr Ward, however, did not consider the matter in that light; he assumed his hypothesis as true, guided, it would seem, by nothing but the opinion, that a centre of uniform motion must somewhere exist, and pleased with the simplicity thus introduced into astronomical calculation. It is, indeed, remarkable, as Montucla has observed, how little the most enlightened astronomers of that time seem to have studied or understood the laws discovered by Kepler. Riccioli, of whom we are just about to speak, enumerates all the suppositions that had been laid down concerning the velocities of the planets, but makes no mention of their describing equal areas in equal times round the sun. Even Cassini, great as he was in astronomy, cannot be entirely exempted from this censure.

Riccioli, a good observer, and a learned and diligent compiler, has collected all that was known in astronomy about the middle of the seventeenth century, in a voluminous work, the *New Almagest*. Without much originality, he was a very useful author, having had, as the historian of astronomy remarks, the courage and the industry to read, to know, and to abridge every thing. He was, nevertheless, an enemy to the Copernican system, and has the discredit of having measured the evidence for and against that system, not by the weight, but by the number of the arguments. The pains which he took to prop the falling edifice of deferents and epicycles, added to his misapprehending and depreciating the discoveries of Kepler, subject him to the reproach of having neither the genius to discover truth, nor the good sense to distinguish it when discovered. He was, however, a priest and a jesuit; he had seen the fate of Galileo; and his errors may have arisen from want of courage, more than from want of discernment.

Of the phenomena which the telescope in the hands of Galileo had made known, the most paradoxical were those exhibited by Saturn; sometimes attended by two globes, one on each side, without any relative motion, but which would, at stated times, disappear for a while, and leave the planet single, like the other heavenly bodies. Nearly forty years had elapsed, without any farther insight into these mysterious appearances, when Huygens began to examine the heavens with telescopes of his own construction, better and more powerful than any which had yet been employed. The two globes that had appeared insulated, were now seen connected by a circular and luminous belt, going quite round the planet. At last, it was found that all these appearances resulted from a broad ring surrounding Saturn, and seen obliquely from the earth. The gradual manner in which this truth unfolded itself is very interesting, and has been given with the detail that it deserves by Huygens, in his *Systema Saturninum.*

The attention which Huygens had paid to the ring of Saturn, led him to the discovery of a satellite of the same planet. His telescopes were not powerful enough to discover more of them than one; he believed, indeed, that there were no more, and that the number of the planets now discovered was complete. The reasoning by which he convinced himself, is a proof how slowly men are cured of their prejudices, even with the best talents and the best information. The planets, primary and secondary, thus made up twelve, the double of six, the first of the perfect numbers. In 1671, however, Cassini discovered another satellite, and afterwards three more, making five in all, which the more perfect telescopes of Dr Herschell have lately augmented to seven.

To the genius of Huygens astronomy is indebted for an addition to its apparatus, hardly less essential than the quadrant and the telescope. An accurate measure of time is of use even in the ordinary business of life, but to the astronomer is infinitely valuable. The dates of his observations, and an accurate estimate of the time elapsed between them, is necessary, in order to make them lead to any useful consequences. Besides this, the only way of measuring with accuracy those arches in the heavens, which extend from east to west, or which are parallel to the equator, depends on the earth's rotation, because such an arch bears the same proportion to the entire circumference of a circle, that the time of its passage under the meridian bears to an entire day. The reckoning of time thus furnishes the best measure of position, as determined by arches parallel to the equator, whether on the earth or in the heavens.

Though the pendulum afforded a measure of time, in itself of the greatest exactness, the means of continuing its motion, without disturbing the time of its vibrations, was yet required to be found, and this, by means of the clock, Huygens contrived most ingeniously to effect. Each vibration of the pendulum, by means of an arm at right angles to it, allows the tooth of a wheel to escape, the wheel being put in motion by a weight. The wheel is so contrived, that the force with which it acts is just sufficient to restore to the pendulum the motion which it had lost by the resistance of the air, and the friction at the centre of motion. Thus the motion of the clock is continued without any diminution of its uniformity, for any length of time.

The telescope had not yet served astronomy in all the capacities in which it could be useful. Huygens, of whose inventive genius the history of science has so much to record, applied it to the measurement of small angles, forming it into the instrument which has since been called a micrometer. By introducing into the focus of the telescope a round aperture of a given size, he contrived to measure the angle which that aperture subtended to the eye, by observing the time that a star placed near the equator required to traverse it. When the angle subtended by any other object in the telescope was to be measured, he introduced into the focus a thin piece of metal, just sufficient to cover the object in the focus. The proportion of the breadth of this plate, to the diameter of the aperture formerly measured, gave the angle subtended by the image in the focus of the telescope. This contrivance is described in the *Systema Saturninum*, at the end.

The telescope has farther contributed materially to the accuracy of astronomical observation, by its application to instruments used for measuring, not merely small angles, but angles of any magnitude whatever. The telescope here comes in place of the plain sights with which the index or *allidad* of an instrument used to be directed to an object, and this substitution has been accompanied with two advantages. The disk of a star is never so well defined to the naked eye as it is in the telescope. Besides, in using plain sights, the eye adapts itself to the farther off of the two, in order that its aperture may be distinctly seen. Whenever this adjustment is made, the object seen through the aperture necessarily appears indistinct to the eye, which is then adapted to a near object. This circumstance produces an uncertainty in all such observations, which, by the use of the telescope, is entirely removed.

But the greatest advantage arises from the magnifying power of the telescope, from which it follows, that what is a mere point to the naked eye, is an extended line which can be divided into a great number of parts when seen through the former. The best eye, when not aided by glasses, is not able to perceive an object which subtends an angle less than half a minute, or thirty seconds. When the index of a quadrant, therefore, is directed by the naked eye to any point in the heavens, we cannot be sure that it is nearer than half a minute on either side of that point. But when we direct the axis of a telescope, which magnifies thirty times, to the same object, we are sure that it is within the thirtieth part of half a minute, that is, within one second of the point aimed at. Thus the accuracy *cæteris paribus* is proportional to the magnifying power.

The application of the telescope, however, to astronomical instruments, was not introduced without opposition. Hevelius of Dantzic, the greatest observer who had been since Tycho Brahé, who had furnished his observatory with the best and largest instruments, and who was familiar with the use of the telescope, strenuously maintained the superiority of the plain sights. His principal argument was founded on this,—that, in plain sights, the line of collimation is determined in its position by two fixed points at a considerable distance from one another, viz. the centres of the two apertures of the sights, so that it remains invariable with respect to the index.

In the case of the telescope there was one fixed point, the intersection of the wires in the focus of the eye-glass; but Hevelius did not think that the other point, viz. the optical centre of the object-glass, was equally well defined. This doubt, however, might have been removed by a direct appeal to experiment, or to angles actually measured on the ground, first by an instrument, and then by trigonometrical operations. From thence it would soon have been discovered, that the centre of a lens is in fact a point defined more accurately than can be done by any mechanical construction.

This method of deciding the question was not resorted to. Hevelius and Hooke had a very serious controversy concerning it, in which the advantage remained with the latter. It should have been observed that the French astronomer, Picard, was the first who employed instruments furnished with telescopic sights, about the year 1665. It appears, however, that Gascoigne, an English gentleman who fell at the battle of Marston-moor in 1644, had anticipated the French astronomer in this invention, but that it had remained entirely unknown. He had also anticipated the invention of the micrometer. The vast additional accuracy thus given to instruments formed a new era in the history of astronomical observations.

Though Galileo had discovered the satellites of Jupiter, their times of revolution, and even some of their inequalities, it yet remained to define their motions with precision, and to construct tables for calculating their places. This task was performed by the elder Cassini, who was invited from Italy, his native country, by Louis the Fourteenth, and settled in France in 1669. His tables of the satellites had been published at Bologna three years before, and he continued to improve them, by a series of observations made in the observatory at Paris, with great diligence and accuracy.

The theory of the motions of these small bodies is a research of great difficulty, and had been attempted by many astronomers before Cassini, with very little success. The planes of the orbits, their inclinations to the orbit of Jupiter, and the lines in which they intersected that orbit, were all to be determined, as well as the times of revolution, and the distances of each from its primary. Add to this, that it is only in a few points of their orbits that they can be observed with advantage. The best are at the times of immersion into the shadow of Jupiter, and emersion from it. The same excellent astronomer discovered four satellites of Saturn, in addition to that already observed by Huygens. He also discovered the rotation of Jupiter and of Mars upon their axes.

The constant attention bestowed on the eclipses of the satellites of Jupiter, made an inequality be remarked in the periods of their return, which seemed to depend on the position of the earth relatively to Jupiter and the sun, and not, as the inequalities of that sort might have been expected to do, on the place of Jupiter in his orbit. From the opposition of Jupiter to the sun, till the conjunction, it was found, that the observed emersion of the satellites from the shadow fell more and more behind the computed; the differences amounting near the conjunction to about fourteen minutes. When, after the conjunction, the immersions were observed, an acceleration was remarked just equal to the former retardation, so that, at the opposition, the eclipse happened fourteen minutes sooner than by the calculation.

The first person who offered an explanation of these facts was Olaus Roemer, a Danish astronomer. He observed that the increase of the retardation corresponded nearly to the increase of the earth's distance from Jupiter, and conversely, the acceleration to the diminution of that distance. Hence it occurred to him, that it was to the time which light requires to traverse those distances that the whole series of phenomena was to be ascribed. This explanation was so simple and satisfactory, that it was readily received.

Though Roemer was the first who communicated this explanation to the world, yet it seems certain that it had before occurred to Cassini, and that he was prevented from making it known by a consideration which does him great honour. The explanation which the motion of light afforded, seemed not to be consistent with two circumstances involved in the phenomenon. If such was the cause of the alternate acceleration and retardation above described, why was it observed only in the eclipses of the first satellite, and not in those of the other three? This difficulty appeared so great to Cassini, that he suppressed the explanation which he would otherwise have given.

The other difficulty occurred to Maraldi. Why did not an equation or allowance of the same kind arise from the position of Jupiter, with respect to his aphelion, for, all other things being the same, his distance from the earth must be greater, as he was nearer to that point of his orbit? Both these difficulties have since been completely removed. If the aforesaid inequality was not for some time observed in any satellite but the first, it was only because the motions of the first are the most regular, and were the soonest understood, but it now appears that the same equation belongs to all the satellites. The solution of Maraldi's difficulty is similar; for the quantity of what is called the equation of the light, is now known to be affected by Jupiter's place in his orbit.

Thus, every thing conspires to prove the reality of the motion of light, so singular on account of the immensity of the velocity, and the smallness of the bodies to which it is communicated.

5. Establishment of Academies, &c.

About the middle of the seventeenth century were formed those associations of scientific men, which, under the appellation of Academies or Philosophical Societies, have contributed so much to the advancement of knowledge in Europe. The *Academia del Cimento* of Florence, founded in 1651, carried in its name the impression of the new philosophy. It was in the country of Galileo where the first institution for the prosecution of experimental knowledge might be expected to arise, and the monuments which it has left behind it will ever create regret for the shortness of its duration.

England soon after showed the same example. It has been already remarked, that, during the civil wars, a number of learned and scientific men sought, in the retirement of Oxford, an asylum from the troubles to which the country was then a prey. They had met as early as 1645; most of them were attached to the royal cause; and after the restoration of Charles the Second, they were incorporated by a royal charter in 1662.

The first idea of this institution seems to have been suggested by the writings of Bacon, who, in recommending the use of experiment, had severely censured the schools, colleges, and academies of his own time, as adverse to the advancement of knowledge; * and, in the *Nova Atlantis*, had given a

* "In moribus et institutis scholarum, academiarum, collegiorum, et similium conventuum, quæ doctorum hominum sedibus, et eruditionis culturæ destinata sunt, omnia progressui scientiarum adversa inveniuntur. Lectiones enim et exercitia ita sunt disposita, ut aliud a consuetis haud facile cuiquam in mentem veniat cogitare, aut contemplari. Si vero unus aut alter fortasse judicii libertate uti sustinuerit, is sibi soli hanc operam imponere possit; ab aliorum autem consortio nihil capiet utilitatis. Sin et hoc toleraverit, tamen in capessenda fortuna industriam hanc et magnanimitatem sibi non levi impedimento fore experietur. Studia enim hominum in ejusmodi locis, in quorundam auctorum scripta, veluti in carceres, conclusa sunt; a quibus si quis dissentiat, continuo ut homo turbidus et rerum novarum cupidus corripitur. *In artibus autem et scientiis tanquam in metalli-fodinis omnia novis operibus et ulterioribus progressibus circumstrepere debent.*"—*Nov. Org.* Lib. i. Aph. 90.

It would be gratifying to be able to observe, that the uni-

most interesting sketch of the form of a society, directed to scientific improvement. In Germany, the *Academia Naturæ Curiosorum* dates its commencement from 1652, and the historian of that institution ascribes the spirit which produced it to the writings of the philosopher just named. These examples, and a feeling that the union and co-operation of numbers was necessary to the progress of experimental philosophy, operated still more extensively. The Royal Academy of Sciences at Paris was founded in 1666, in the reign of Louis the Fourteenth, and during the administration of Colbert. The Institute of Bologna in Italy belongs nearly to the same period; but almost all the other philosophical associations, of which there are now so many, had their beginning in the eighteenth century.

Frequent communication of ideas, and a regular

---

versities of Europe had contributed to the renovation of science. The fact is otherwise;—they were often the fastnesses from which prejudice and error were latest of being expelled. They joined in persecuting the reformers of science. It has been seen, that the masters of the University of Paris were angry with Galileo for the experiments on the descent of bodies. Even the University of Oxford brought on itself the indelible disgrace of persecuting, in Friar Bacon, the first man who appears to have had a distinct view of the means by which the knowledge of the laws of nature must be acquired.

method of keeping up such communication, are evidently essential to works in which great labour and industry are to be employed, and to which much time must necessarily be devoted; when the philosopher must not always sit quietly in his cabinet, but must examine nature with his own eyes, and be present in the work-shop of the mechanic, or the laboratory of the chemist. These operations are facilitated by the institutions now referred to, which, therefore, are of more importance to the physical sciences than to the other branches of knowledge. They who cultivate the former are also fewer in number, and being, of course, farther separated, are less apt to meet together in the common intercourse of the world. The historian, the critic, the poet, finds everywhere men who can enter in some degree at least into his pursuits, who can appreciate his merit, and derive pleasure from his writings or his conversation. The mathematician, the astronomer, the mechanician, sees few men who have much sympathy with his pursuits, or who do not look with indifference on the objects which he pursues. The *world*, to him, consists of a few individuals, by the censures or approbation of whom the public opinion must be finally determined; with them it is material that he should have more frequent intercourse than could be obtained by casual rencounter; and he feels that the society of men engaged in pursuits similar to his own, is a necessary *stimulus* to his exertions. Add to this, that such societies become centres in which information concerning facts is collected from all quarters. For all these reasons, the greatest benefit has resulted from the scientific institutions which, since the middle of the seventeenth century, have become so numerous in Europe.

The Royal Society of London is an association of men, who, without salaries or appointment from Government, defray, by private contribution, the expence of their meetings, and of their publications. This last is another important service, which a society so constituted renders to science.

The demand of the public for memoirs in mathematics and natural philosophy, many of them perhaps profound and difficult, is not sufficiently great to defray the expence of publication, if they come forward separately and unconnected with one another. In a collective state they are much more likely to draw the attention of the public; the form in which they appear is the most convenient both for the reader and the author; and if, after all, the sale of the work is unequal to the expence, the deficiency is made up from the funds of the society. An institution of this kind, therefore, is a patriotic and disinterested association of the lovers of science, who engage not only to employ themselves in discovery, but, by private contribution, to defray the expence of scientific publications.

The Academy of Sciences in Paris was not exactly an institution of the same kind. It consisted of three classes of members, one of which, the *Pensionnaires*, twenty in number, had salaries paid by Government, and were bound in their turns to furnish the meetings with scientific memoirs, and each of them also, at the beginning of every year, was expected to give an account of the work in which he was to be employed. This institution has been of incredible advantage to science. To detach a number of ingenious men from every thing but scientific pursuits; to deliver them alike from the embarrassments of poverty or the temptations of wealth; to give them a place and station in society the most respectable and independent, is to remove every impediment, and to add every stimulus to exertion. To this institution, accordingly, operating upon a people of great genius, and indefatigable activity of mind, we are to ascribe that superiority in the mathematical sciences, which, for the last seventy years, has been so conspicuous.

The establishment of astronomical observatories, as national or royal works, is connected in Europe with the institution of scientific or philosophical societies. The necessity of the former was, indeed, even more apparent than that of the latter. A science, which has the heavenly bodies for its objects, ought, as far as possible, to be exempted from the vicissitudes of the earth. As it gains strength but slowly, and requires ages to complete its discoveries, the plan of observation must not be limited by the life of the individual who pursues it, but must be followed out in the same place, year after year, to an unlimited extent. A perception of this truth, however indistinct, seems, from the earliest times, to have suggested the utility of observatories, to those sovereigns who patronised astronomy, whether they looked to that science for real or imaginary instruction. The circle of Osymandias is the subject of one of the most ancient traditions in science, and has preserved the name of a prince which otherwise would have been entirely unknown. A building, dedicated to astronomy, made a conspicuous part of the magnificent establishment of the school of Alexandria. During the middle ages, in the course of the migrations of science toward the east, sumptuous buildings, furnished with astronomical instruments, rose successively in the plains of Mesopotamia, and among the mountains of Tartary. An observatory in the gardens of the Caliph of Bagdat contained a quadrant of fifteen cubits * in radius, and a sextant of forty. † Instruments of a still larger size distinguished the observatory of Sa-

* Twenty-two feet three inches.

† Sixty feet five inches.

marcande, and the accounts would seem incredible, if the ruins of Benares did not, at this moment, attest the reality of similar constructions.

On the revival of letters in Europe, establishments of the same kind were the first decisive indications of a taste for science. We have seen the magnificent observatory on which Tycho expended his private fortune, and employed the munificence of his patron, become a sad memorial (after the signal services which it had rendered to astronomy) of the instability of whatever depends on individual greatness. The observatories at Paris and London were secured from a similar fate, by being made national establishments, where a succession of astronomers were to devote themselves to the study of the heavens. The observatory at Paris was begun in 1667, and that at Greenwich in 1675. In the first of these, La Hire and Cassini, in the second, Flamstead and Halley, are at the head of a series of successors, who have done honour to their respective nations. If there be in Britain any establishment, in the success and conduct of which the nation has reason to boast, it is that of the Royal Observatory, which, in spite of a climate which so continually tries the patience, and so often disappoints the hopes of the astronomer, has furnished a greater number of observations to be completely relied on, than all the rest of Europe put together, and afforded the *data* for those tables, in which the French mathematicians have expressed, with such accuracy, the past, the present, and the future condition of the heavens.

6. Figure and Magnitude of the Earth.

The progress made during the seventeenth century, in ascertaining the magnitude and figure of the earth, is particularly connected with the establishments which we have just been considering. Concerning the figure of the earth, no accurate information was derived from antiquity, if we except that of the mathematical principle on which it was to be determined. The measurement of an arch of the meridian was attempted by Eratosthenes of Alexandria, in perfect conformity with that principle, but by means very inadequate to the importance and difficulty of the problem. By measuring the sun's distance from the zenith of Alexandria, on the solstitial day, and by knowing, as he thought he did, that, on the same day, the sun was exactly in the zenith of Syené, he found the distance in the heavens between the parallels of those places to be 7° 12′, or a 50th part of the circumference of a great circle. Supposing, then, that Alexandria and Syené were in the same meridian, nothing more was required than to find the distance between them, which, when multiplied by 50, would give the circumference of the globe. The manner in which this was attempted by Eratosthenes is quite characteristic of the infant state of the arts of experiment and observation. He took no trouble to ascertain whether Alexandria and Syené were due north and south of one another: the truth is, that the latter is considerably east of the former, so that, though their horizontal distance had been accurately known, a considerable reduction would have been necessary, on account of the distance of the one from the meridian of the other. It does not appear, however, that Eratosthenes was at any more pains to ascertain the distance than the bearing of the two places. He assumed the former just as it was commonly estimated; and, indeed, it appears that the distance was not measured till long afterwards, when it was done by the command of Nero.

It was in this way that the ancients made observations and experiments; the mathematical principles might be perfectly understood, but the method of obtaining accurate data for the application of those principles was not a subject of attention. The *power* of resolving the problem was the main object; and the actual solution was a matter of very inferior importance. The slowness with which the art of making accurate experiments and observations has been matured, and the great distance it has kept behind theory, is a remarkable fact in the history of the physical sciences. It has been remarked, that mathematicians had found out the area of the circle, and calculated its circumference to more than a hundred places of decimals, before artists had divided an arch into minutes of a degree; and that many excellent treatises had been written on the properties of curves, before a straight line had been drawn of any considerable length, or measured with any tolerable exactness, on the surface of the globe. *

The next measurement on record is that of the astronomers of Almamon, in the plains of Mesopotamia, and the manner of conducting the operation appears to have been far more accurate than that of the Greek philosophers; but, from a want of knowledge of the *measures* employed, it has conveyed no information to posterity.

The first arch of the meridian measured in modern times with an accuracy any way corresponding to the difficulty of the problem, was by Snellius, a Dutch mathematician, who has given an account of it in a volume which he calls *Eratosthenes Batavus*, published in 1617. The arch was between Bergen-op-zoom and Alkmaar; its amplitude was 1° 11′ 30″, and the distance was determined by a series of triangles, depending on a base line carefully measured. The length of the degree that resulted was 55,021 toises, which, as was afterwards found, is considerably too small. Certain errors were dis-

* Edinburgh Review, Vol. V. p. 391.

covered, and when they were corrected, the degree came out 57,033 toises, which is not far from the truth. The corrections were made by Snellius himself, who measured his base over again, and also the angles of the triangles. He died, however, before he could publish the result. Muschenbroek, who calculated the whole anew from his papers, came to the conclusion just mentioned, which, of course, was not known till long after the time when the measure was executed. No advantage, accordingly, was derived to the world from this measurement till its value was lost in that of other measurements still more accurately conducted.

A computation which, for the time, deserves considerable praise, is that of Norwood, in 1635, who measured the distance between London and York, taking the bearings as he proceeded along the road, and reducing all to the direction of the meridian, and to the horizontal plane. The difference of latitude he found, by observation of the solstices, to be 2° 28′, and from that and his measured distance, he concluded the degree to be 367,176 feet English, or 57,800 toises. This has been found to be a near approximation; yet his method was not capable of great accuracy, nor did he always execute it in the best manner. "Sometimes," says he, "I measured, sometimes I *paced*, and I believe I am within a *scantling* of the truth."

Fernel, a French physician, measured with a

wheel from Paris to Amiens, which are nearly in the same meridian, and he determined the degree from thence to be 56,746 French toises; a result which falls short of the truth, though not very considerably.

These investigations, it is plain, could not but leave considerable uncertainty with respect to the magnitude of the earth. The Academy of Sciences became interested in the question, and the measurement of an arch in the meridian was undertaken under its auspices, and executed by the Abbé Picard, already known for his skill in the operations of practical geometry. He followed a method similar to that of Snellius, according to which, the distance between Amiens and Malvoisin was found from a series of triangles, and a base of $5663\frac{1}{6}$ toises. He determined the difference of latitude by means of a zenith sector of ten feet radius, and found it to be $1° 22' 55''$. The whole distance was 78,850 toises, whence the degree came out 57,060 toises. This was the first measurement of a degree of the meridian, on which perfect reliance could be placed.

Hitherto no doubt had been entertained of the spherical figure of the earth, and, of consequence, of the equality of all the degrees of the meridian, so that, if one was known, the whole circumference was determined. Men, with the precipitation which they so often manifest, of assuming, without sufficient evidence, the conclusion which appears most simple, were no sooner satisfied that the earth was round, than they supposed it to be truly spherical. An observation soon occurred, which gave reason to suspect, that much more must be done before its figure or its magnitude were completely ascertained.

With a view of observing the sun's altitude in the vicinity of the equator, where the distance from the zenith being inconsiderable, the effects of refraction must be of small account, it was agreed, by the same academy, to send an astronomer, M. Richer, to make observations at the island of Cayenne, in South America.

Richer observed the solstitial altitude of the sun at that place in 1672, and found the distance of the tropics to be 46° 57′ 4″; and, therefore, the obliquity of the ecliptic 23° 28′ 32″, agreeing almost precisely with the determination of Cassini.

The most remarkable circumstance, however, which occurred in the course of this voyage, was, that the clock, though furnished with a pendulum of the same length which vibrated seconds at Paris, was found, at Cayenne, to lose two minutes and a half a day nearly. This created great astonishment in France, especially after the accuracy of it was confirmed by the observations of Varin and Deshayes, who, some years afterwards, visited different places on the coast of Africa and America, near the line, and found the necessity of shortening the pendulum, to make it vibrate seconds in those latitudes. The first explanation of this remarkable phenomenon was given by Newton, in the third book of his *Principia*, published in 1687, where it is deduced as a necessary consequence of the earth's rotation on its axis, and of the centrifugal force thence arising. That force changes both the direction and the intensity of gravity, giving to the earth an oblate spheroidal figure, more elevated at the equator than the poles, and making bodies fall, and pendulums vibrate, more slowly in low than in high latitudes.

This solution, however, did not, any more than the book in which it was contained, make its way very readily into France. The first explanation of the retardation of the pendulum, which was received there, was given by Huygens in 1690. Huygens deduced it also from the centrifugal force, arising from the earth's rotation, and the view which he took was simpler, though much less accurate than that of Newton. It had, indeed, the simplicity which often arises from neglecting one of the essential conditions of a problem ; but it was nevertheless ingenious, and involved a very accurate knowledge of the nature of centrifugal force. I am thus brought to touch on a subject which belongs properly to the second part of this Dissertation, for which the fuller discussion of it must of course be reserved.

SECTION V.

OPTICS.

1. OPTICAL KNOWLEDGE OF THE ANCIENTS.

ON account of the rectilineal propagation of light, the phenomena of optics are easily expressed in the form of mathematical propositions, and seem, as it were, spontaneously to offer themselves to the study of geometers. Euclid perceiving this affinity, began to apply the science which he had already cultivated with so much success, to explain the laws of vision, before a similar attempt had been made with respect to any other branch of terrestrial physics, and at least fifty years before the researches of Archimedes had placed mechanics among the number of the mathematical sciences.

In the treatise ascribed to Euclid, there are, however, only two physical principles which have completely stood the test of subsequent improvement. The first of these is the proposition just referred to, that a point in any object is seen in the direction of a straight line drawn from the eye to that point; and the second is, that when a point in an object is seen by reflection from a polished surface, the lines drawn from the eye and from the object to the point whence the reflection is made, are equally inclined to the reflecting surface. These propositions are assumed as true; they were, no doubt, known before the time of Euclid, and it is supposed that the discovery of them was the work of the Platonic school. The first of them is the foundation of Optics *proper*, or the theory of vision by direct light; the second is the foundation of *Catoptrics*, or the theory of vision by reflected light. Dioptrics, or vision by refracted light, had not yet become an object of attention.

Two other principles which Euclid adopted as postulates in his demonstrations, have not met with the same entire confirmation from experiment, and are, indeed, true only in certain cases, and not universally, as he supposed. The first of these is, that we judge of the magnitude of an object altogether by the magnitude of the optical angle, or the angle which it subtends at the eye. It is true that this angle is an important element in that judgment, and Euclid, by discovering this, came into the possession of a valuable truth; but by a species of sophistry, very congenial to the human mind, he extended the principle too far, and supposed it to be the only circumstance which determines our judgment of visible magnitude. It is, indeed, the only measure which we are furnished with directly by the eye itself; but there are few cases in which we form our estimate without first appealing to the commentary afforded by the sensations of touch, or the corrections derived from our own motion.

Another principle, laid down by the same geometer, is in circumstances nearly similar to the preceding. According to it, the place of any point of an object seen by reflection, is always the intersection of the reflected ray, with the perpendicular drawn from that point to the reflecting surface. The proof offered is obscure and defective; the proposition, however, is true, of plane specula always, and of spherical as far as Euclid's investigations extended, that is, while the rays fall on the speculum with no great obliquity. His assumption, therefore, did not affect the truth of his conclusions, though it would have been a very unsafe guide in more general investigations. The book is in many other respects imperfect, the reasoning often unsound, and the whole hardly worthy of the great geometer whose name it bears. There is, however, no doubt that Euclid wrote on the subject of optics, and many have supposed that this treatise is a careless extract, or an unskilful abridgment of the original work.

Antiquity furnished another mathematical treatise on optics, that of the astronomer Ptolemy. This treatise, though known in the middle ages, and quoted by Roger Bacon, had disappeared, and was supposed to be entirely lost, till within these few years, when a manuscript on optics, professing to be the work of Ptolemy, and to be translated from the Arabic, was found in the King's Library at Paris. The most valuable part of this work is that which relates to refraction, from whence it appears that many experiments had been made on that subject, and the angles of incidence and refraction, for different transparent substances, observed with so much accuracy, that the same ratio very nearly of the sines of these angles, from air into water, or into glass, is obtained from Ptolemy's numbers, which the repeated experiments of later times have shown to be true. The work, however, in the state in which it now appears, is very obscure, the reasoning often deficient in accuracy, and the mathematical part much less perfect than might have been expected. Modern writers, presuming partly on the reputation of Ptolemy, and partly guided by the authority of Roger Bacon, had ascribed to this treatise more merit than it appears to possess; and, of consequence, had allowed less to the Arabian author Alhazen, who comes next in the order of time, than of right belongs to him. Montucla, on the authority of Bacon, says, that Ptolemy ascribed the increase of the apparent magnitude of the heavenly bodies near the horizon, to the greater distance at which they are supposed to be, on account of the number of intervening objects across which they are seen. Ptolemy's explanation, however, as stated by Delambre, * from the manuscript just mentioned, is quite different from this, and amounts to no more than the vague and unsatisfactory remark, that an observer looks at the bodies near the zenith in a constrained posture, and in a situation to which the eye is not accustomed. The former explanation, therefore, given by Alhazen, but supposed to have been borrowed from Ptolemy, must now be returned to its right owner. It is the best explanation yet known.

These are the only mathematical treatises on optics of any consideration which the ancients have transmitted to us; † but many metaphysical spe-

---

* *Connaissance des Tems*, 1816, p. 245, &c. The glimpses of truth, not destined to be fully discovered till many ages afterwards, which are found in the writings of the ancients, are always interesting. Ptolemy distinguishes what has since been called the *virtual focus*, which takes place in certain cases of reflection from spherical specula. He remarks, that colours are confounded by the rapidity of motion, and gives the instance of a wheel painted with different colours, and turned quickly round.

† Another Greek treatise on optics, that of *Heliodorus of Larissa*, has been preserved, and was first published by Erasmus Bartholinus at Paris, in 1657. It is a superficial work, which, to a good deal of obscure and unsound metaphysics, adds the demonstration of a few very obvious truths. The author holds the opinion, that vision is performed by the emission of something from the eyes; and the reason which he assigns is, that the eyes are convex, and more

culations on light and vision are to be found in the writings of the philosophers. Aristotle defined light much as he had defined motion; *the act or energy of a transparent body, in as much as it is transparent.* The reason for calling light an act of a transparent body is, that, though a body may be transparent in power or capacity, it does not become actually transparent but by means of light. Light brings the transparency into action; it is, therefore, the act of a transparent body. In such miserable puerilities did the genius of this great man exhaust itself, owing to the unfortunate direction in which his researches were carried on.

In his farther speculations concerning light, he denied it to be a substance; and his argument contains a singular mixture of the ingenious and the absurd. The time, he says, in which light spreads from one place to another is infinitely small, so that light has a velocity which is infinitely great. Now, bodies move with a velocity inversely as the quantities of matter which they contain; light, therefore, cannot contain any matter,

---

adapted to emit than to receive. His metaphysics may be judged of from this specimen. He has not been made mention of by any ancient author, and the time when he wrote is unknown. As he quotes, however, the writings of Ptolemy and Hero, he must have been later than the first century.

that is, it cannot be material. * That the velocity of light was infinitely great, seemed to him to follow from this, that its progress, estimated either in the direction of north and south, or of east and west, appeared to be instantaneous. In the opinion of the Platonists, and of the greater part of the ancients, vision was performed by means of certain rays which proceeded from the eye to the object, though they did not become the instruments of conveying sensations to the mind, but in consequence of the presence of light. In this theory, we can now see nothing but a rude and hasty attempt to assimilate the sense of sight to that of touch, without inquiring sufficiently into the particular characters of either.

Epicurus, and the philosophers of his school, as we learn from Lucretius, entertained more correct notions of vision, though they were still far from the truth. They conceived vision to be performed in consequence of certain *simulacra*, or images continually thrown off from the surfaces of bodies, and entering the eye. This was the substitute in their philosophy for rays of light, and had at least the merit of representing that which is the medium of vision, or which forms the communication be-

---

* The truth of the mathematical proposition, that $\frac{1}{\text{inf.}}=0$, was perceived by Aristotle. A strong intellect is always visible in the midst of his greatest errors.

tween the eye and external objects, as something proceeding from the latter. The idea of *simulacra*, or *spectra*, flying off continually from the surfaces of bodies, and entering the eye, was perhaps as near an approach to the true theory of vision as could be made before the structure of the eye was understood.

In the arts connected with optics, the ancients had made some progress. They were sufficiently acquainted with the laws of reflection to construct mirrors both plane and spherical. They made them also conical; and it appears from Plutarch, that the fire of Vesta, when extinguished, was not permitted to be rekindled but by the rays of the sun, which were condensed by a conical speculum of copper. The mirrors with which Archimedes set fire to the Roman gallies have been subjects of much discussion, and the fact was long disbelieved, on the ground of being physically impossible. The experiments of Kircher and Buffon showed that this impossibility was entirely imaginary, and that the effect ascribed to the *specula* of the Greek geometer might be produced without much difficulty. There remains now no doubt of their reality. A passage from Aristophanes * gives reason to believe that, in his time, lenses of glass were used for burning, by collecting the rays of

* In Nubibus, Act. 2, sc. 1. v. 20.

the sun; but in a matter that concerns the history of science, the authority of a comic poet and a satirist would not deserve much attention, if it were not confirmed by more sober testimony. Pliny, speaking of rock crystal, * says that a globe or ball of that substance was sometimes used by the physicians for collecting the rays of the sun, in order to perform the operation of cautery. In another passage, he mentions the power of a glass globe filled with water, to produce a strong heat when exposed to the rays of the sun, and expresses his surprise that the water itself should all the while remain quite cold.

With respect to the power of glasses to magnify objects seen through them, or to render such objects more distinct, the ancients appear to have observed ill, and to have reasoned worse. "Literæ quamvis minutæ et obscuræ per vitream pilam aqua plenam majores clarioresque cernuntur. Sidera ampliora per nubem adspicienti videntur: *quia acies nostra in humido labitur, nec apprehendere quod vult fideliter potest.*" † This passage, and the speculations concerning the rainbow in the same place, when they are considered as containing the opinions of some of the most able and best informed men of antiquity, must be admitted to

* Hist. Nat. Lib. 37. cap. 10.

† Seneca, Nat. Quest. Lib. i. cap. 6.

mark, in a very striking manner, the infancy of the physical sciences.

2. From Alhazen to Kepler.

An interval of nearly a thousand years divided Ptolemy from Alhazen, who, in the history of optical discovery, appears as his immediate successor. This ingenious Arabian lived in the eleventh century, and his merit can be more fairly, and will be more highly appreciated, now that the work of his predecessor has become known. The merit of his book on Optics was always admitted, but he was supposed to have borrowed much from Ptolemy, without acknowledging it; and the prejudices entertained in favour of a Greek author, especially of one who had been for so many years a legislator in science, gave a false impression, both of the genius and the integrity of his modern rival. The work of Alhazen is, nevertheless, in many respects, superior to that of Ptolemy, and in nothing more than in the geometry which it employs. The problem known by his name, to find the point in a spherical speculum, at which a ray coming from one given point shall be reflected to another given point, is very well resolved in his book, though a problem of so much difficulty, that Montucla hazards the opinion, that no Arabian geometer was ever equal to the solution of it.* It is now certain, however, that the solution, from whatever quarter it came, was not borrowed from Ptolemy, in whose work no mention is made of any such question; and it may very well be doubted, whether, had this problem been proposed to him, the Greek geometer would have appeared to as much advantage as the Arabian.

The account which the latter gives of the augmentation of the diameters of the heavenly bodies near the horizon has been already mentioned. He treated also of the refraction of light by transparent bodies, and particularly of the atmospheric refraction, but not with the precision of Ptolemy, whose optical treatise Delambre seems to think it probable that he had never seen. The anatomical structure of the eye was known to him; concerning the uses of the different parts he had only conjectures to offer; but on seeing single with two eyes, he

* Barrow, in his 9th lecture, says of this Problem, that it may truly be called δυσμήχανον, as hardly any one more difficult had then been attempted by geometers. He adds, that, after trying the analysis in many different ways, he had found nothing preferable to the solution of Alhazen, which he therefore gives, only freed from the prolixness and obscurity with which the original is chargeable. *Lectiones Opticæ*, Sect. 9. p. 65. A very elegant solution of the same problem is given by Simson, at the end of his Conic Sections.

made this very important remark, that, when corresponding parts of the retina are affected, we perceive but one image.

Prolixity and want of method are the faults of Alhazen. Vitello, * a learned Pole, commented on his works, and has very much improved their method and arrangement in a treatise published in 1270. He has also treated more fully of the subject of refraction, and reduced the results of his experiments into the form of a table exhibiting the angles of refraction corresponding to the angles of incidence, which he had tried in water and glass. It was not, however, till long after this period that the law which connects these angles was discovered. The cause of refraction appeared to him to be the resistance which the rays suffer in passing into the denser medium of water or glass, and one can see in his reasoning an obscure idea of the resolution of forces. He also treats of the rainbow, and remarks, that the altitudes of the sun and bow together always amount to 42 degrees. He next considers the structure of the eye, of which he has given a tolerably accurate description, and proves, as Alhazen had before done, † that vision is not performed by the emission of rays from the eye.

---

* The name of this author is commonly written *Vitellio*. He may be supposed to have known best the orthography of his own name.

† Alhazen, Opt. Lib. 1.

Roger Bacon, distinguished for pursuing the path of true philosophy in the midst of an age of ignorance and error, belongs to the same period; and applied to the study of optics with peculiar diligence. It does not appear, however, that he added much to the discoveries of Alhazen and Ptolemy, with whose writings, particularly those of the former, he seems to have been well acquainted. In some things he was much behind the Arabian optician, as he supposed with the ancients that vision is performed by rays emitted from the eye. It must, however, be allowed, that the arguments employed on both sides of this question are so weak and inconclusive, as very much to diminish the merit of being right, and the demerit of being wrong. What is most to the credit of Bacon, is the near approach he appears to have made to the knowledge of lenses, and their use in assisting vision. Alhazen had remarked, that small objects, letters, for instance, viewed through a segment of a glass sphere, were seen magnified, and that it is the larger segment which magnifies the most. The spherical segment was supposed to be laid with its base on the letters, or other minute objects which were to be viewed. Bacon recommends the smaller segment, and observes, that the greater, though it magnify more, places the object farther off than its natural position, while the other brings it nearer. This shows sufficiently, that he knew how to trace the progress of the rays of light through a spherical transparent body, and understood, what was the thing least obvious, how to determine the place of the image. Smith, in his Optics, endeavours to show, that these conclusions were purely theoretical, and that Roger Bacon had never made any experiments with such glasses, notwithstanding that he speaks as if he had done so. * This severe remark proceeds on some slight inaccuracy in Bacon's description, which, however, does not seem sufficient to authorize so harsh a conclusion. The probability appears rather to be, as Molineux supposed, that Bacon had made experiments with such glasses, and was both practically and theoretically acquainted with their properties. At the same time, it must be acknowledged, that his credulity on many points, and his fondness for the marvellous, which, with every respect for his talents, it is impossible to deny, take something away from the force of his testimony, except when it is very expressly given. However that may be in the present case, it is probable, that the knowledge of the true properties of these glasses, whether it was theoretical or practical, may have had a share in introducing the use of lenses, and in the invention of spectacles, which took place not long after.

It would be desirable to ascertain the exact pe-

* Smith's Optics, Vol. II. Remarks, § 76.

riod of an invention of such singular utility as this last; one that diffuses its advantages so widely, and that contributes so much to the solace and comfort of old age, by protecting the most intellectual of the senses against the general progress of decay. In the obscurity of a dark age, careless about recording discoveries of which it knew not the principle or the value, a few faint traces and imperfect indications serve only to point out certain limits within which the thing sought for is contained. Seeking for the origin of a discovery, is like seeking for the source of a river where innumerable streams have claims to the honour, between which it is impossible to decide, and where the only thing that can be known with certainty is the boundary by which they are all circumscribed. The reader will find the evidence concerning the invention of spectacles very fully discussed in Smith's Optics, from which the most probable conclusion is, that the date goes back to the year 1313, and cannot with any certainty be traced farther. *

The lapse of more than two hundred years brings us down to Maurolycus, and to an age when men of science ceased to be so thinly scattered over the wastes of time. Maurolycus, whose knowledge of the pure mathematics has been already mention-

* Smith's Optics, Vol. II. Remarks, § 75.

ed, was distinguished for his skill in optics. He was acquainted with the crystalline lens, and conceived that its office is to transmit to the optic nerve the *species* of external objects; and in this process he does not consider the retina as any way concerned. This theory, though so imperfect, led him nevertheless to form a right judgment of the defects of short-sighted and long-sighted eyes. In one of his first works, Theoremata de Lumine et Umbra, he also gives an accurate solution of a question proposed by Aristotle, viz. why the light of the sun, admitted through a small hole, and received on a plane at a certain distance from it, always illuminates a round space, whatever be the figure of the hole itself, whereas, through a large aperture, the illuminated space has the figure of the aperture. To conceive the reason of this, suppose that the figure of the hole is a triangle; it is plain that at each angle the illuminated space will be terminated by a circular arch of which the centre corresponds to the angular point, and the radius to the angle subtended by the sun's semidiameter. Thus the illuminated space is rounded off at the angles; and when the hole is so small that the size of those roundings bears a large proportion to the distance of their centres, the figure comes near to a circle, and may be to appearance quite round. This is the true solution, and the same with that of Maurolycus. The same author appears also to have observed the caustic curve formed by reflection from a concave speculum

A considerable step in optical discovery was made at this time by Baptista Porta, a Neapolitan, who invented the *Camera Obscura*, about the year 1560, and described it in a work, entitled *Magia Naturalis*. The light was admitted through a small hole in the window-shutter of a dark room, and gave an inverted picture of the objects from which it proceeded, on the opposite wall. A lens was not employed in the first construction of this apparatus, but was afterwards used; and Porta went so far as to consider how the effect might be produced without inversion. He appears to have been a man of great ingenuity; and though much of the *Magia Naturalis* is directed to frivolous objects, it indicates a great familiarity with experiment and observation. It is remarkable, that we find mention made in it of the reflection of cold by a speculum,* an experiment which, of late, has drawn so much attention, and has been supposed to be so entirely new. The cold was perceived by making the focus fall on the eye, which, in the absence of the thermometer, was, perhaps, the best measure of small variations

* Magia Naturalis, Lib. 17. cap. 4. p. 583. Amsterdam edit. 1664

of temperature. Porta's book was extremely popular; and when we find it quickly translated into Italian, French, Spanish, and Arabic, we see how much the love of science was now excited, and what effects the art of printing was now beginning to produce. Baptista Porta was a man of fortune, and his house was so much the resort of the curious and learned at Naples, that it awakened the jealousy with which the court of Rome watched the progress of improvement. How grievous it is to observe the head of the Christian church, in that and the succeeding age, like the *Anarch old* in Milton, reigning in the midst of darkness, and complaining of the encroachments which the realm of light was continually making on his ancient empire!

The constitution of the eye, and the functions of the different parts of which it consists, were not yet fully understood. Maurolycus had nearly discovered the secret, and it was but a thin, though, to him, an impenetrable veil, which still concealed one important part of the truth. This veil was drawn aside by the Neapolitan philosopher; but the complete discovery of the truth was left to Kepler, who, to the glory of finding out the true laws of the planetary system, added that of first analyzing the whole scheme of nature in the structure of the eye. He perceived the exact resemblance of this organ to the *dark chamber*, the rays entering the pupil being collected by the crystalline lens, and the other humours of the eye, into *foci*, which paint on the *retina* the inverted images of external objects. By another step of the process, to which our analysis can never be expected to extend, the mind perceives the images thus formed, and refers them at the same time to things without.

It seemed a great difficulty, that, though the images be inverted, the objects are seen erect; but when it is considered that each point in the object is seen in the direction of the line in which the light passes from it to the retina, through the centre of the eye, it will appear that the upright position of the object is a necessary consequence of this arrangement.

Kepler's discovery is explained in his *Paralipomena in Vitellionem*,* (Remarks on the Optics of Vitello,) a work of great genius, abounding with new and enlarged views, though mixed occasionally with some unsound and visionary speculations. This book appeared in 1604. In the next article we shall have occasion to return to the consideration of other parts of Kepler's optical discoveries.

---

* Caput 5. de Modo Visionis.

3. From Kepler to the commencement of Newton's Optical Discoveries.

The rainbow had, from the earliest times, been an object of interest with those who bestowed attention on optical appearances, but it is much too complicated a phenomenon to be easily explained. In general, however, it was understood to arise from light reflected by the drops of rain falling from a cloud opposite to the sun. The difficulty seemed to be how to account for the colour, which is never produced in white light, such as that of the sun, by mere reflection. Maurolycus advanced a considerable step when he supposed that the light enters the drop, and acquires colour by refraction; but in tracing the course of the ray he was quite bewildered. Others supposed the refraction and the colour to be the effect of one drop, and the reflection of another; so that two refractions and one reflection were employed, but in such a manner as to be still very remote from the truth.

Antonio de Dominis, Archbishop of Spalatro, had the good fortune to fall upon the true explanation. Having placed a bottle of water opposite to the sun, and a little above his eye, he saw a beam of light issue from the under side of the bottle, which acquired different colours, in the same order, and with the same brilliancy as in the rainbow, when the bottle was a little raised or depressed. From comparing all the circumstances, he perceived that the rays had entered the bottle, and that, after two refractions from the convex part, and a reflection from the concave, they were returned to the eye tinged with different colours, according to the angle at which the ray had entered. The rays that gave the same colour made the same angle with the surface, and hence all the drops that gave the same colour must be arranged in a circle, the centre of which was the point in the cloud opposite to the sun. This, though not a complete theory of the rainbow, and though it left a great deal to occupy the attention, first of Descartes, and afterwards of Newton, was perfectly just, and carried the explanation as far as the principles then understood allowed it to go. The discovery itself may be considered as an anomaly in science, as it is one of a very refined and subtle nature, made by a man who has given no other indication of much scientific sagacity or acuteness. In many things his writings show great ignorance of principles of optics well known in his own time, so that Boscovich, an excellent judge in such matters, has said of him, "homo opticarum rerum, supra id quod patiatur ea ætas imperitissimus."

The book containing this discovery was published in 1611.*

A discovery of the same period, but somewhat earlier, will always be considered as among the most remarkable in the whole circle of human knowledge. It is the invention of the telescope, the work in which, (by following unconsciously the plan of nature in the formation of the eye,) man has come the nearest to the construction of a new organ of sense. For this great invention, in its original form, we are indebted to accident, or to the trials of men who had little knowledge of the principles of the science on which they were conferring so great a favour. A series of scientific improvements, continued for more than two hundred years, has continually added to the perfection of this noble instrument, and has almost entitled science to consider the telescope as its own production.

It will readily be believed, that the origin of such an invention has been abundantly inquired into. The result, however, as is usual in such cases, has not been quite satisfactory; and all that is known with certainty is, that the honour belongs to the town of Middleburgh in Zealand, and that the

* De Radiis Lucis in Vitris perspectivis et Iride.—Venetiis, in 4to.

date is between the last ten years of the sixteenth century, and the first ten of the seventeenth. Two different workmen belonging to that town, Zachariah Jans, and John Lapprey, have testimonies in their favour between which it is difficult to decide; the former goes back to 1590, the latter comes down to about 1610. It is not of much consequence to settle the priority in a matter which is purely accidental; yet one would not wish to forget or mistake the names of men whom even chance had rendered so great benefactors to science. What we know with certainty is, that the account of the effect produced by this new combination of glasses being carried to Galileo in 1610, led that great philosopher to the construction of the telescope, and to the interesting discoveries already enumerated. By what principle he was guided to the combination, which consists of one convex and one concave lens, he has not explained, and we cannot now exactly ascertain. He had no doubt observed, that a convex lens, such as was common in spectacles, formed images of objects, which were distinctly seen when thrown on a wall or on a screen. He might observe also, that if the image, instead of falling on the screen, were made to fall on the eye, the vision was confused and indistinct. In the trials to remedy this indistinctness, by means of another glass, it would be found that a concave lens succeeded when placed before the eye, the eye itself being also a little more advanced than the screen had been.

This instrument, though very imperfect, compared with those which have been since constructed, gave so much satisfaction, that it remained long without any material improvement. Descartes, whose treatise on Optics was written near thirty years after the invention of the telescope, makes no mention of any but such as is composed of a convex object-glass, and a concave eye-glass. The theory of it, indeed, was given by Kepler in his Dioptrics, (1611,) when he also pointed out the astronomical telescope, or that which is composed of two convex lenses, and inverts the objects. He did not, however, construct a telescope of that kind, which appears to have been first done by *Scheiner*, who has given an account of it in the *Rosa Ursina*, (in 1650,) quoted by Montucla. *

After the invention of the telescope, that of the microscope was easy ; and it is also to Galileo that we are indebted for this instrument, which discovers an immensity on the one side of man, scarcely less wonderful than that which the telescope discovers on the other. The extension and divisibility of matter are thus rendered to the natural philosopher almost as unlimited as the extension and the divisibility of space are to the geometer.

---

* Vol. II. p. 234, 2d edit.

The theory of the telescope, now become the main object in optical science, required that the law of refraction should, if possible, be accurately ascertained. This had not yet been effected, and Kepler, whose *Dioptrics* was the most perfect treatise on refraction which had yet appeared, had been unable to determine the general principle which connects the angles of incidence and refraction. In the case of glass, he had found by experiment, that those angles, when small, are nearly in the ratio of three to two, and on this hypothesis he had found the focus of a double convex lens, when the curvature of both sides is equal, to be the centre of curvature of the side turned toward the object,—a proposition which is known to coincide with experiment. From the same approximation, he derived other conclusions, which were found useful in practice, in the cases where the angles just mentioned were very small.

The discovery of the true law of refraction was the work of Snellius, the same mathematician whose labours concerning the figure of the earth were before mentioned. In order to express this law, he supposed a perpendicular to the refracting surface, at the point where the refraction is made, and also another line parallel to this perpendicular at any given distance from it. The refracted ray, as it proceeds, will meet this parallel, and the incident ray is supposed to be produced, till it do so likewise. Now, the general truth which Snellius found to hold, whatever was the position of the incident ray, is, that the segments of the refracted ray and of the incident ray, intercepted by these parallels, had always the same ratio to one another. If either of the *media* were changed, that through which the incident ray, or that through which the refracted ray passed, this ratio would be changed, but while the *media* remained the same, the ratio continued unalterable. It is seldom that a general truth is seen at first under the most simple aspect : this law admits of being more simply expressed, for, in the triangle formed by the two segments of the rays, and by the parallel which they intersect, the said segments have the same ratio with the sines of the opposite angles, that is, with the sines of the angles of incidence and refraction. The law, therefore, comes to this, that, in the refraction of light, by the same medium, the sine of the angle of incidence has to the sine of the angle of refraction always the same ratio. This last simplification did not occur to Snellius ; it is the work of Descartes, and was first given in his Dioptrics, in 1637, where no mention is made of Snellius, and the law of refraction appears as the discovery of the author. This naturally gave rise to heavy charges against the candour and integrity of the French philosopher. The work of Snellius had never been published, and the author himself was dead ; but the proposition just referred to had been communicated to his friends, and had been taught by his countryman, Professor Hortensius, in his lectures. There is no doubt, therefore, that the discovery was first made by Snellius, but whether Descartes derived it from him, or was himself the second discoverer, remains undecided. The question is one of those, where a man's conduct in a particular situation can only be rightly interpreted from his general character and behaviour. If Descartes had been uniformly fair and candid in his intercourse with others, one would have rejected with disdain a suspicion of the kind just mentioned. But the truth is, that he appears throughout a jealous and suspicious man, always inclined to depress and conceal the merit of others. In speaking of the inventor of the telescope, he has told minutely all that is due to accident, but has passed carefully over all that proceeded from design, and has incurred the reproach of relating the origin of that instrument, without mentioning the name of Galileo. In the same manner, he omits to speak of the discoveries of Kepler, so nearly connected with his own; and in treating of the rainbow, he has made no mention of Antonio de Dominis. It is impossible that all this should not produce an unfavourable impression, and hence it is, that even the warmest admirers of Descartes do not pretend that his conduct toward Snellius can be completely justified.

Descartes would have conceived his philosophy to be disgraced if it had borrowed any general principle from experience, and he therefore derived, or affected to derive, the law of refraction from reasoning or from theory. In this reasoning, there were so many arbitrary suppositions concerning the nature of light, and the action of transparent bodies, that no confidence can be placed in the conclusions deduced from it. It is indeed quite evident, that, independently of experiment, Descartes himself could have put no trust in it, and it is impossible not to feel how much more it would have been for the credit of that philosopher to have fairly confessed that the knowledge of the law was from experiment, and that the business of theory was to deduce from thence some inferences with respect to the constitution of light and of transparent bodies. This I conceive to be the true method of philosophizing, but it is the reverse of that which Descartes pursued on all occasions.

The weakness of his reasoning was perceived and attacked by Fermat, who, at the same time, was not very fortunate in the theory which he proposed to substitute for that of his rival. The latter had laid it down as certain, that light, of which he supposed the velocity infinite, or the propagation instantaneous, meets with less obstruction in dense than in rare bodies, for which reason, it is refracted toward the perpendicular, in passing from the latter into the former. This seemed to Fermat a very improbable supposition, and he conceived the contrary to be true, viz. that light in rare bodies has less obstruction, and moves with greater velocity than in dense bodies. On this supposition, and appealing, not to physical, but to final causes, Fermat imagined to himself that he could deduce the true law of refraction. He conceived it to be a fact that light moves always between two points, so as to go from the one to the other in the least time possible. Hence, in order to pass from a given point in a rarer medium where it moves faster, to a given point in a denser medium where it moves slower, so that the time may be a *minimum*, it must continue longer in the former medium than if it held a rectilineal course, and the bending of its path, on entering the latter, will therefore be toward the perpendicular. On instituting the calculus, according to his own doctrine of *maxima* and *minima*, Fermat found, to his surprise, that the path of the ray must be such, that the sines of the angles of incidence and refraction have a constant ratio to one another. Thus did these philosophers, setting out from suppositions entirely contrary, and following routes which only agreed in being quite unphilosophical and arbitrary, arrive, by a very unexpected coincidence, at the same conclusion. Fermat could no longer deny the law of refraction, as laid down by Descartes, but he was less than ever disposed to admit the justness of his reasoning.

Descartes proceeded from this to a problem which, though suggested by optical considerations, was purely geometrical, and in which his researches were completely successful. It was well known, that, in the ordinary cases of refraction by spherical and other surfaces, the rays are not collected into one point, but have their foci spread over a certain surface, the sections of which are the curves called caustic curves, and that the focus of opticians is only a point in this surface, where the rays are more condensed, and, of course, the illumination more intense than in other parts of it. It is plain, however, that if refraction is to be employed, either for the purpose of producing light or heat, it would be a great advantage to have all the rays which come from the same point of an object united accurately, after refraction, in the same point of the image. This gave rise to an inquiry into the figure which the superficies, separating two transparent *media* of different refracting powers, must have, in order that all the rays diverging from a given point might, by refraction at the said superficies, be made to converge to another given point.* The problem was resolved by Descartes in its full extent; and he proved, that the curves, proper for

* Cartesii Dioptrices, cap. 8vum; Geometria, lib. 2dus.

generating such superficies by their revolution, are all comprehended under one general character, viz. that there are always two given points, from which, if straight lines be drawn to any point in the curve, the one of these, *plus* or *minus* that which has a given ratio to the other, is equal to a given line.

It is evident, when the given ratio here mentioned is a ratio of equality, that the curve is a conic section, and the two given points its two foci. The curves, in general, are of the fourth or the second order, and have been distinguished by the name of the ovals of Descartes.

From this very ingenious investigation no practical result of advantage in the construction of lenses has been derived. The mechanical difficulties of working a superficies into any figure but a spherical one are so great, that, notwithstanding all the efforts of Descartes himself, and of many of his followers, they have never been overcome, so that the great improvements in optical instruments have arisen in a quarter entirely different.

Descartes gave also a full explanation of the rainbow,* as far as colour was not concerned, a part of the problem which remained for Newton to resolve. The path of the ray was traced, and the angles of the incident ray, with that which emerges after two refractions and one reflection, was accurately deter-

* Meteorum, cap. 8vum.

mined. Descartes paid little attention to those who had gone before him, and, as already remarked, never once mentioned the Archbishop of Spalatro. Like Aristotle, he seems to have formed the design of cutting off the memory of all his predecessors, but the invention of printing had made this a far more hopeless undertaking than it was in the days of the Greek philosopher.

After the publication of the Dioptrics of Descartes, in 1637, a considerable interval took place, during which optics, and indeed science in general, made but little progress, till the Optica Promota of James Gregory, in 1663, seemed to put them again in motion. The author of this work, a profound and inventive geometer, had applied diligently to the study of optics and the improvement of optical instruments. The Optica Promota embraced several new inquiries concerning the illumination and distinctness of the images formed in the foci of lenses, and contained an account of the Reflecting Telescope still known by the name of its author. The consideration which suggested this instrument was the imperfection of the images formed by spherical lenses, in consequence of which, they are not in plane, but in curved surfaces. The desire of removing this imperfection led Gregory to substitute reflection for refraction in the construction of telescopes; and by this means, while he was seeking to remedy a small evil, he provided the means of avoiding a much greater one, with which he was not yet acquainted, viz. that which arises from the unequal refrangibility of light. The attention of Newton was about the same time drawn to the same object, but with a perfect knowledge of the defect which he wanted to remove. Gregory thought it necessary that the specula should be of a parabolic figure; and the execution proved so difficult, that the instrument, during his own life, was never brought to any perfection. The specula were afterwards constructed of the ordinary spherical form, and the Gregorian telescope, till the time of Dr Herschel, was more in use than the Newtonian.

Gregory was Professor of Mathematics at St Andrews, and afterwards for a short time at Edinburgh. His writings strongly mark the imperfect intercourse which subsisted at that time between this country and the Continent. Though the Optics of Descartes had been published twenty-five years, Gregory had not heard of the discovery of the law of refraction, and had found it out only by his own efforts;—happy in being able, by the fertility of his genius, to supply the defects of an insulated and remote situation.

A course of lectures on optics, delivered at Cambridge in 1668, by Dr Barrow, and published in the year following, treated of all the more difficult questions which had occurred in that state of the science, with the acuteness and depth which are found in all the writings of that geometer. This work contains some new views in optics, and a great deal of profound mathematical discussion.

About this time Grimaldi, a learned jesuit, the companion of Riccioli, in his astronomical labours, made known some optical phenomena which had hitherto escaped observation. They respected the action of bodies on light, and when compared with reflection and refraction, might be called, in the language of Bacon's philosophy, *crepuscular* instances, indicating an action of the same kind, but much weaker and less perceptible. Having stretched a hair across a sun-beam, admitted through a hole in the window-shutter of a dark chamber, he was surprised to find the shadow much larger than the natural divergence of the rays could have led him to expect. Other facts of the same kind made known the general law of the *diffraction* or *inflection* of light, and showed that the rays are acted on by bodies, and turned out of their rectilineal course, even when not in contact, but at a measurable distance from the surfaces or edges of such bodies. Grimaldi gave an account of those facts in a treatise printed at Bologna in 1665.*

Optics, as indeed all the branches of natural philosophy, have great obligations to Huygens. The former was among the first scientific objects which

---

* Physico-Mathesis de Lumine, Coloribus, &c. in 4to.

occupied his mind; and his Dioptrics, though a posthumous work, is most of it the composition of his early youth. It is written with great perspicuity and precision, and is said to have been a favourite book with Newton himself. Though beginning from the first elements, it contains a full developement of the matters of greatest difficulty in the construction of telescopes, particularly in what concerns the indistinctness arising from the imperfect foci into which rays are united by spherical lenses; and rules are deduced for constructing telescopes, which, though of different sizes, shall have the same degree of distinctness, illumination, &c. Huygens was besides a practical optician; he polished lenses, and constructed telescopes with his own hands, and some of his object-glasses were of the enormous focal distance of 130 feet. To his Dioptrics is added a valuable treatise De Formandis Vitris.

In the history of optics, particular attention is due to his theory of light, which was first communicated to the Academy of Sciences of Paris, in 1678, and afterwards published, with enlargements, in 1690.*

Light, according to this ingenious system, consists in certain undulations communicated by luminous bodies to the etherial fluid which fills all space. This fluid is composed of the most subtle

* Traité de la Lumière. Leyd. 1690.

matter, is highly elastic, and the undulations are propagated through it with great velocity in spherical superficies proceeding from a centre. Light, in this view of it, differs from that of the Cartesian system, which is supposed to be without elasticity, and to convey impressions instantaneously, as a staff does from the object it touches to the hand which holds it.

It is not, however, in this general view, that the ingenuity of the theory appears, but in its application to explain the equality of the angles of incidence and reflection; and, most of all, the constant ratio which subsists between the sines of the angles of incidence and of refraction. Few things are to be met with more simple and beautiful than this last application of the theory; but that which is most remarkable of all is, the use made of it to explain the double refraction of Iceland crystal. This crystal, which is no other than the calcareous spar of mineralogists, has not only the property of refracting light in the usual manner of glass, water, and other transparent bodies, but it has also another power of refraction, by which even the rays falling perpendicularly on the surface of the crystal are turned out of their course, so that a double image is formed of all objects seen through these crystals. This property belongs not only to calcareous spar, but, in a greater or less degree, to all substances which are both crystallized and transparent.

The common refraction is explained by Huygens, on the supposition, that the undulations in the luminous fluid are propagated in the form of spherical waves. The double refraction is explained on the supposition, that the undulations of light, in passing through the calcareous spar, assume a spheroidal form; and this hypothesis, though it does not apply with the same simplicity as the former, yet admits of such precision, that a proportion of the axes of the spheroids may be assigned, which will account for the precise quantity of the extraordinary refraction, and for all the phenomena dependent on it, which Huygens had studied with great care, and had reduced to the smallest number of general facts. That these spheroidal undulations actually exist, he would, after all, be a bold theorist who should affirm; but that the supposition of their existence is an accurate expression of the phenomena of double refraction, cannot be doubted. When one enunciates the hypothesis of the spheroidal undulations, he, in fact, expresses in a single sentence all the phenomena of double refraction. The hypothesis is therefore the means of representing these phenomena, and the laws which they obey, to the imagination or the understanding, and there is, perhaps, no theory in optics, and but very few in natural philosophy, of which more can be said. Theory, therefore, in this instance, is merely to be regarded as the expression of a general law, and in that light, I think, it is considered by La Place.

To carry the theory of Huygens farther, and to render it quite satisfactory, a reason ought to be assigned why the undulations of the luminous fluid are spheroidal in the case of crystals, and spherical in all other cases. This would be to render the generalization more complete; and till that is done, and a connection clearly established between the structure of crystallized bodies, and the property of double refraction, the theory will remain imperfect. The attention which at present is given to this most singular and interesting branch of optics, and the great number of new phenomena observed and classed under the head of the *Polarisation of Light*, make it almost certain that this object will be either speedily accomplished, or that science has here reached one of the immoveable barriers by which the circle of human knowledge is to be for ever circumscribed.

END OF PART I.

DISSERTATION.

PART II.

DISSERTATION.

PART II.

In the former part of this sketch, the history of each division of the sciences was continued without interruption, from the beginning to the end. During the period, however, on which I am now to enter, the advancement of knowledge has been so rapid, and marked by such distinct steps, that several pauses or resting-places occur of which it may be advisable to take advantage. Were the history of any particular science to be continued for the whole of the busy interval which this second part embraces, it would leave the other sciences too far behind; and would make it difficult to perceive the mutual action by which they have so much assisted the progress of one another. Considering some sort of subdivision, therefore, as necessary, and observing, in the interval which extends from the first of Newton's discoveries to the year 1818, three different conditions of the Physico-Mathematical sciences, well marked and distinguished by great improvements, I have divided the above interval into three corresponding parts. The first of these, reaching from the commencement of Newton's discoveries in 1663, to a little beyond his death, or to 1730, may be denominated, from the men who impressed on it its peculiar character, *the period of Newton and Leibnitz.* The second, which, for a similar reason, I call that of *Euler and D'Alembert*, may be regarded as extending from 1730 to 1780; and the third, that of *Lagrange and Laplace*, from 1780 to 1818.

PERIOD FIRST.

SECTION I.

THE NEW GEOMETRY.

THE seventeenth century, which had advanced with such spirit and success in combating prejudice, detecting error, and establishing truth, was destined to conclude with the most splendid series of philosophical discoveries yet recorded in the history of letters. It was about to witness, in succession, the invention of Fluxions, the discovery of the Composition of Light, and of the Principle of Universal Gravitation,—all three within a period of little more than twenty years, and all three the work of the same individual. It is to the first of these that our attention at present is to be particularly directed.

The notion of Infinite Quantity had, as we have already seen, been for some time introduced into Geometry, and having become a subject of reasoning and calculation, had, in many instances, after facilitating the process of both, led to conclusions from which, as if by magic, the idea of infinity had entirely disappeared, and left the geometer or the algebraist in possession of valuable propositions, in which were involved no magnitudes but such as could be readily exhibited. The discovery of such results had increased both the interest and extent of mathematical investigation.

It was in this state of the sciences, that Newton began his mathematical studies, and, after a very short interval, his mathematical discoveries. * The book, next to the elements, which was put into his hands, was Wallis's *Arithmetic of Infinites*, a work well fitted for suggesting new views in geometry, and calling into activity the powers of mathematical invention. Wallis had effected the quadrature of all those curves in which the value of one of the co-ordinates can be expressed in terms of the other, without involving either fractional or negative exponents. Beyond this point neither his researches, nor those of any other geometer, had yet reached, and from this point the discoveries of Newton began. The Savilian Professor had himself been extremely desirous to advance into the new region, where, among other great objects, the quadrature of the circle must necessarily be contained, and he made a very noble effort to pass the barrier by which the undiscovered country ap-

* He entered at Trinity College, Cambridge, in June 1660. The date of his first discoveries is about 1663.

peared to be defended. He saw plainly, that if the equations of the curves which he had squared were ranged in a regular series, from the simpler to the more complex, their areas would constitute another corresponding series, the terms of which were all known. He farther remarked, that, in the first of these series, the equation to the circle itself might be introduced, and would occupy the middle place between the first and second terms of the series, or between an equation to a straight line and an equation to the common parabola. He concluded, therefore, that if, in the second series, he could interpolate a term in the middle, between its first and second terms, this term must necessarily be no other than the area of the circle. But when he proceeded to pursue this very refined and philosophical idea, he was not so fortunate; and his attempt toward the requisite interpolation, though it did not entirely fail, and made known a curious property of the area of the circle, did not lead to an indefinite quadrature of that curve.* Newton was much more judicious and successful in his attempt. Proceeding on the same general prin-

* The interpolation of Wallis failed, because he did not employ literal or general exponents. His theorem, expressing the area of the entire circle by a fraction, of which the numerator and denominator are each the continued product of a certain series of numbers, is a remarkable anticipation of some of Euler's discoveries, *Calc. Int.* Tom. I. cap. 8.

ciple with Wallis, as he himself tells us, the simple view which he took of the areas already computed, and of the terms of which each consisted, enabled him to discover the law which was common to them all, and under which the expression for the area of the circle, as well as of innumerable other curves, must needs be comprehended. In the case of the circle, as in all those where a fractional exponent appeared, the area was exhibited in the form of an infinite series.

The problem of the quadrature of the circle, and of so many other curves, being thus resolved, Newton immediately remarked, that the law of these series was, with a small alteration, the law for the series of terms which expresses the root of any binomial quantity whatsoever. Thus he was put in possession of another valuable discovery, the Binomial Theorem, and at the same time perceived that this last was in reality, in the order of things, placed before the other, and afforded a much easier access to such quadratures than the method of interpolation, which, though the first road, appeared now neither to be the easiest nor the most direct.

It is but rarely that we can lay hold with certainty of the thread by which genius has been guided in its first discoveries. Here we are proceeding on the authority of the author himself, for, in a letter to Oldenburg, * Secretary of the Royal

* Commercium Epistolicum, Art. 55.

Society of London, he has entered into considerable detail on this subject, adding, (so ready are the steps of invention to be forgotten,) that the facts would have entirely escaped his memory, if he had not been reminded of them by some notes which he had made at the time, and which he had accidentally fallen on. The whole of the letter just referred to is one of the most valuable documents to be found in the history of invention.

In all this, however, nothing occurs from which it can be inferred that the method of fluxions had yet occurred to the inventor. His discovery consisted in the method of reducing the value of y, the ordinate of a curve, into an infinite series of the integer powers of x the abscissa, by division, or the extraction of roots, that is, by the Binomial Theorem; after which, the part of the area belonging to each term could be assigned by the arithmetic of infinites, or other methods already known. He has assured us himself, however, that the great principle of the new geometry was known to him, and applied to investigation as early as 1665 or 1666. * Independently of that authority, we also know, on the testimony of Barrow, that soon after the period just mentioned, there was put into his hands by Newton a manuscript treatise, † the same

* Quadrature of Curves, Introduction.

† Com. Epist. No. I. II. III. &c.

which was afterwards published under the title of *Analysis per Æquationes Numero Terminorum Infinitas*, in which, though the instrument of investigation is nothing else than infinite series, the principle of fluxions, if not fully explained, is at least distinctly pointed out. Barrow strongly exhorted his young friend to publish this treasure to the world; but the modesty of the author, of which the excess, if not culpable, was certainly in the present instance very unfortunate, prevented his compliance. All this was previous to the year 1669; the treatise itself was not published till 1711, more than forty years after it was written.

For a long time, therefore, the discoveries of Newton were only known to his friends, and the first work in which he communicated any thing to the world on the subject of fluxions was in the first edition of the Principia, in 1687, in the second Lemma of the second book, to which, in the disputes that have since arisen about the invention of the new analysis, reference has been so often made. The principle of the fluxionary calculus was there pointed out, but nothing appeared that indicated the peculiar algorithm, or the new notation, which is so essential to that calculus. About this Newton had yet given no information; and it was only from the second volume of Wallis's Works, in 1693, that it became known to the world.* It

* Wallis says, that he had inserted in the English edition

was no less than ten years after this, in 1704, that Newton himself first published a work on the new calculus, his Quadrature of Curves, more than twenty-eight years after it was written.

These discoveries, however, even before the press was employed as their vehicle, could not remain altogether unknown in a country where the mathematical sciences were cultivated with zeal and diligence. Barrow, to whom they were first made known by the author himself, communicated them to Oldenburg, the Secretary of the Royal Society, who had a very extensive correspondence all over Europe. By him the series for the quadrature of the circle were made known to James Gregory, in Scotland, who had occupied himself very much with the same subject. They were also communicated to Leibnitz in Germany, who had become

---

of his book, published in 1685, several extracts from Newton's Letters, "*Omissis multis aliis inibi notatu dignis, eo quod speraverim clarissimum virum voluisse tum illa, tum alia quæ apud ipsum premit edidisse. Cum vero illud nondum fecerit libet eorum nonnulla hic attingere ne pereant.*" Among these last is an account of the fluxionary notation, according to which the fluxions of flowing quantities are distinguished by points, and also of certain applications of this new algorithm, extracted from two letters of Newton, written in 1792.—*Opera,* Tom. II. p. 390, &c.—There is no evidence of this notation having existed earlier than that date, though it be highly probable that it did.

acquainted with Oldenburg, in a visit which he made to England in 1673. At the time of that visit, Leibnitz was but little conversant with the mathematics; but having afterwards devoted his great talents to the study of that science, he was soon in a condition to make new discoveries. He invented a method of squaring the circle, by transforming it into another curve of an equal area, but having the ordinate expressed by a rational fraction of the abscissa, so that its area could be found by the methods already known. In this way he discovered the series, so remarkable for its simplicity, which gives the value of a circular arch in terms of the tangent. This series he communicated to Oldenburg in 1674, and received from him in return an account of the progress made by Newton and Gregory in the invention of series. In 1676, Newton described his method of quadratures at the request of Oldenburg, in order that it might be transmitted to Leibnitz in the two letters already mentioned, as of such value by recording the views which guided that great geometer in his earliest, and some of his most important discoveries. The method of fluxions is not communicated in these letters; nor are the principles of it in any way suggested; though there are, in the last letter, two sentences in transposed characters, which ascertain that Newton was then in possession of that method, and employed in speaking of it the same language in which it was afterwards made known. In the following year, Leibnitz, in a letter to Oldenburg, introduces differentials, and the methods of his calculus for the first time. This letter,* which is very important, clearly proves that the author was then in full possession of the principles of his calculus; and had even invented the algorithm and notation.

From these facts, and they are all that bear directly on the question concerning the invention of the infinitesimal analysis, if they be fairly and dispassionately examined, I think that no doubt can remain, that Newton was the first inventor of that analysis, which he called by the name of Fluxions; but that, in the communications made by him, or his friends, to Leibnitz, there was nothing that could convey any idea of the principle on which that analysis was founded, or of the algorithm which it involved. The things stated were merely results; and though some of those relating to the tangents of curves might show the author to be in possession of a method of investigation different from infinite series, yet they afforded no indication of the nature of that method, or the principles on which it proceeded.

In what manner Newton's communications in the two letters already referred to, may have acted in stimulating the curiosity and extending or even

* Commercium Epistolicum, No. 66.

directing the views of such a man as Leibnitz, I shall not presume to decide, (nor even, if such effect be admitted, will it take from the originality of his discoveries;) but that in the authenticated communications which took place between these philosophers, there was nothing which could make known the nature of the fluxionary calculus, I consider as a fact most fully established.

Of the new or infinitesimal analysis, we are, therefore, to consider Newton as the first inventor, Leibnitz as the second; his discovery, though posterior in time, having been made independently of the other, and having no less claim to originality. It had the advantage also of being first made known to the world; an account of it, and of its peculiar algorithm, having been inserted in the first volume of the *Acta Eruditorum*, in 1684. Thus, while Newton's discovery remained a secret, communicated only to a few friends, the geometry of Leibnitz was spreading with great rapidity over the Continent. Two most able coadjutors, the brothers James and John Bernoulli, joined their talents to those of the original inventor, and illustrated the new methods by the solution of a great variety of difficult and interesting problems. The reserve of Newton still kept his countrymen ignorant of his geometrical discoveries, and the first book that appeared in England on the new geometry was that of Craig, who professedly derived his knowledge from the writings of Leibnitz and his friends. Nothing, however, like rivalship or hostility between these inventors had yet appeared ; each seemed willing to admit the originality of the other's discoveries ; and Newton, in the passage of the *Principia* just referred to, gave a highly favourable opinion on the subject of the discoveries of Leibnitz.

The quiet, however, that now prevailed between the English and German philosophers, was clearly of a nature to be easily disturbed. With the English was conviction, and, as we have seen, a well grounded conviction, that the first discovery of the Infinitesimal Analysis was the property of Newton; but the analysis thus discovered was yet unknown to the public, and was in the hands of the inventor and his friends. With the Germans, there was the conviction, also well founded, that the invention of their countryman was perfectly original ; and they had the satisfaction to see his calculus everywhere adopted, and himself considered all over the Continent as the sole inventor. The friends of Newton could not but resist this latter claim, and the friends of Leibnitz, seeing that their master had become the great teacher of the new calculus, could not easily bring themselves to acknowledge that he was not the first discoverer. The tranquillity that existed under such circumstances, if once disturbed, was not likely to be speedily restored.

Accordingly, a remark of Fatio de Duillier, a

mathematician, not otherwise very remarkable, was sufficient to light up a flame which a whole century has been hardly sufficient to extinguish. In a paper on the line of swiftest descent, which he presented to the Royal Society in 1699, was this sentence: "I hold Newton to have been the first inventor of this calculus, and the earliest, by several years, induced by the evidence of facts; and whether Leibnitz, the second inventor, has borrowed any thing from the other, I leave to the judgment of those who have seen the letters and manuscripts of Newton." Leibnitz replied to this charge in the *Leipsic Journal*, without any asperity, simply stating himself to have been, as well as Newton, the inventor; neither contesting nor acknowledging Newton's claim to priority, but asserting his own to the first publication of the calculus.

Not long after this, the publication of Newton's *Quadrature of Curves*, and his *Enumeration of the lines of the third order*, (1705,) afforded the same journalists an opportunity of showing their determination to retort the insinuations of Duillier, and to carry the war into the country of the enemy. After giving a very imperfect synopsis of the first of these books, they add: "*Pro differentiis igitur Leibnitianis D. Newtonus adhibet, semperque adhibuit fluxiones; quæ sunt proxime ut fluentium augmenta, equalibus temporis particulis quam minimis genita; iisque tum in suis Principiis Natu-*

ræ Mathematicis, tum in aliis post editis, eleganter est usus; quemadmodum Honoratus Fabrius in sua Synopsi Geometricâ motuum progressus Cavalierianæ methodo substituit." *

In spite of the politeness and ambiguity of this passage, the most obvious meaning appeared to be, that Newton had been led to the notion of fluxions by the differentials of Leibnitz, just as Honoratus Fabri had been led to substitute the idea of progressive motion for the indivisibles of Cavalieri. A charge so entirely unfounded, so inconsistent with acknowledged facts, and so little consonant to declarations that had formerly come from the same quarter, could not but call forth the indignation of Newton and his friends, especially as it was known, that these journalists spoke the language of Leibnitz and Bernoulli. In that indignation they were perfectly justified; but when the minds of contending parties have become irritated in a certain degree, it often happens that the injustice of one side is retaliated by an equal injustice from the opposite. Accordingly, Keill, who, with more zeal than judgment, undertook the defence of Newton's claims, instead of endeavouring to establish the priority of his discoveries, by an appeal to facts and to dates that could be accurately ascertained, (in which he would have been completely successful,)

* Com. Epist. No. 97. Newtoni *Opera*, Tom. IV. p. 577.

undertook to prove, that the communications of Newton to Leibnitz, were sufficient to put the latter in possession of the principles of the new analysis, after which he had only to substitute the notion of differentials for that of fluxions. In support of a charge which it would have required the clearest and most irresistible evidence to justify, he had, however, nothing to offer but equivocal facts and overstrained arguments, such as could only convince those who were already disposed to believe. They were, accordingly, received as sound reasoning in England, rejected as absurd in Germany, and read with no effect by the mathematicians of France and Italy.

Leibnitz complained of Keill's proceeding to the Royal Society of London, which declined giving judgment, but appointed a commission of its members to draw up a full and detailed report of all the communications which had passed between Newton and Leibnitz, or their friends, on subjects connected with the new analysis, from the time of Collins and Oldenburg to the date of Keill's letter to Sir Hans Sloane in 1711, the same that was now complained of. This report forms what is called the *Commercium Epistolicum ;* it was published by order of the Royal Society the year following, and contains an account of the facts, which, though in the main fair and just, does not give that impression of the impartiality of the reporters, which the circumstances so imperiously demanded. Leibnitz complained of this publication, and alleged, that though nothing might be inserted that was not contained in the original letters, yet certain passages were suppressed which were favourable to his pretensions. He threatened an answer, which, however, never appeared. Some notes were added to the *Commercium*, which contain a good deal of asperity and unsupported insinuation; the *Recensio*, or review of it, inserted in the *Philosophical Transactions* for 1715, though written with ability, is still more liable to the same censure.

In the year (1713) which followed the publication of the *Commercium Epistolicum*, a paragraph was circulated among the mathematicians of Europe, purporting to be the *judgment of a mathematician* on the invention of the new analysis. The author was not named, but was generally understood to be John Bernoulli, of which, indeed, the terms in which Leibnitz speaks of the judgment leaves no room to doubt. Bernoulli was without question well acquainted with the subject in dispute; he was a perfect master of the calculus; he had been one of the great instruments of its advancement, and, except impartiality, possessed every requisite for a judge. Without offence it might be said, that he could scarcely be accounted impartial. He had been a party in all that had happened;—warmly attached as he was to the one side, and greatly exasperated against the other, his temper had been more frequently ruffled, and his passions or prejudices more violently excited, than those of any other individual. With all his abilities, therefore, he was not likely to prove the fairest and most candid judge, in a cause that might almost be considered as his own. His sentence, however, is pronounced in calm and temperate language, and amounts to this, *That there is no reason to believe, that the fluxionary calculus was invented before the differential.* I shall refer to a note * the discussion of the evidence which he points out as the ground of this decision, though the facts already stated might be considered as sufficient to enable the reader to form an opinion on the subject. The friends of Leibnitz hurt their own cause, by attempting to fix on Newton a charge of plagiarism, which was refuted by such a chain of evidence, by so many dates distinctly ascertained, and so many concessions of their own. A candid review of the evidence led to the conviction, that both Newton and Leibnitz were original inventors. When the English mathematicians accused Leibnitz of borrowing from Newton, they were, therefore, going much farther than the evidence authorized them,

* This note having been left in an unfinished state, cannot be presented to the public.—E.

and were mistaking their own partialities for proofs. They maintained what was not true, but what, nevertheless, was not physically impossible, the discovery of Newton being certainly prior to that of Leibnitz. The German mathematicians, on the other hand, when they charged Newton with borrowing from Leibnitz, were maintaining what was not only false, but what involved an impossibility. This is the only part of the dispute, in which any thing that could be construed into *mala fides* can be said to have appeared. I am far, however, from giving it that construction; men of such high character, both for integrity and talents, as Leibnitz and Bernoulli, ought not to be lightly subjected to so cruel an imputation. Partiality, prejudice, and passion, are sufficient to account for much injustice, without a decided intention to do wrong.

In the state of hostility to which matters were now brought, the new analysis itself was had recourse to, as affording to either side abundant means of annoying its adversaries, by an inexhaustible supply of problems, accessible to those alone who were initiated in the doctrines, and who could command the resources of that analysis. The power of resolving such problems, therefore, seemed a test whether this analysis was understood or not. Already some questions of this kind had been proposed in the *Leipsic Journal*, not as defiances, but as exercises in the new geometry. Such was the problem of the *Catenaria*, or the curve, which a chain of uniform weight makes when suspended from two points. This had been proposed by Bernoulli in 1690, and had been resolved by Huygens, Leibnitz, and himself.

A question had been proposed, also, concerning the line of swiftest descent in 1697, or the line along which a body must descend, in order to go from one point to another not perpendicularly under it, in the least time possible. Though a straight line be the shortest distance between two points, it does not necessarily follow, that the descent in that line will be most speedily performed, for, by falling in a curve that has at first a very rapid declivity, the body may acquire in the beginning of its motion so great a velocity, as shall carry it over a long line in less time than it would describe a short one, with a velocity more slowly acquired. This, however, is a problem that belongs to a class of questions of peculiar difficulty; and accordingly it was resolved only by a few of the most distinguished mathematicians. The solutions which appeared within the time prescribed were from Leibnitz, Newton, the two Bernoullis, and M. de l'Hopital. Newton's appeared in the *Philosophical Transactions* without a name; but the author was easily recognized. John Bernoulli, on seeing it, is said to have exclaimed, *Ex ungue leonem!*

The curve that has the property required is the

cycloid; Newton has given the construction, but has not accompanied it with the analysis. He added afterwards the demonstration of a very curious theorem for determining the time of the actual descent. Leibnitz resolved the problem the same day that he received the *programme* in which it was proposed.

The problem of orthogonal trajectories, as it is called, had been long ago proposed in the *Acta Eruditorum*, with an invitation to all who were skilled in the new analysis to attempt the solution. The problem had not, at first, met with the attention it was supposed to deserve, but John Bernoulli having resumed the consideration of it, found out what appeared a very perfect and very general solution; and the question was then (1716) proposed anew by Leibnitz, for the avowed purpose of trying the skill of the English mathematicians. The question is, a system of curves described according to a known law being given, (all the hyperbolas, for instance, that are described between the same asymptotes; or all the parabolas that have the same directrix, and that pass through the same point, &c.) to describe a curve which shall cut them all at right angles. This may be considered as the first defiance professedly aimed at the English mathematicians. The problem was delivered to Newton on his return from the Mint, when he was much fatigued with the business of the day;

he resolved it, however, the same evening, and his solution, though without a name, is given in the *Philosophical Transactions* for 1716. *

This solution, however, only gave rise to new quarrels, for hardly any thing so excellent could come from the one side, that it could meet with the entire approbation of the other. Newton's, indeed, was rather the plan or *projet* of an investigation, than an actual solution; and, in the general view which it took of the question, could hardly provide against all the difficulties that might occur in the application to particular cases. This was what Bernoulli objected to, and affected to treat the solution as of no value. Brook Taylor, Secretary of the Royal Society, and well known as one of the ablest geometers of the time, undertook the defence of it, but concluded with using language very reprehensible, and highly improper to be directed by one man of science against another. Having sufficiently, as he supposed, replied to Bernoulli and his friends, he adds, "If they are not satisfied with the solution, it must be ascribed to *their own ignorance*." † It strongly marks the temper by which both sides were now animated, when a man like Taylor, eminent for profound science, and, in general, very much disposed to do justice to the

* Vol. XXIX. p. 399.

† Eorum imperitiæ tribuendum est.

merits of others, should so forget himself as to reproach with ignorance of the calculus, one of the men who understood it the best, and who had contributed the most to its improvement The irritability and prejudices of Bernoulli admitted of no defence, and he might very well have been accused of viewing the solution of Newton through a medium disturbed by their action ; but to suppose that he was unable to understand it, was an impertinence that could only react on the person who was guilty of it. Bernoulli was not exemplary for his patience, and it will be readily believed, that the incivility of Taylor was sufficiently revenged. It is painful to see men of science engaged in such degrading altercation, and I should be inclined to turn from so disagreeable an object, if the bad effects of the spirit thus excited were not such as must again obtrude themselves on the notice of the reader.

Taylor not long after came forward with an open defiance to the whole Continent, and proposed a problem, *Omnibus geometris non Anglis*,—a problem, of course, which he supposed that the English mathematicians alone were sufficiently enlightened to resolve. He selected one, accordingly, of very considerable difficulty,—the integration of a fluxion of a complicated form ; which, nevertheless, admitted of being done in a very elegant manner, known, I believe, at that time to very few of the English mathematicians, to Cotes, to himself, and, perhaps, one or two more. The selection, nevertheless, was abundantly injudicious; for Bernoulli, as long ago as 1702, had explained the method of integrating this, and such like formulas, both in the *Paris Mémoires* and in the *Leipsic Acts*. The question, accordingly, was no sooner proposed than it was answered in a manner the most clear and satisfactory; so the defiance of Taylor only served to display the address and augment the triumph of his adversary.

The last and most unsuccessful of these challenges was that of Keill, of whose former appearance in this controversy we have already had so much more reason to commend the zeal than the discretion. Among the problems in the mixed mathematics which had excited most attention, and which seemed best calculated to exercise the resources of the new analysis, was the determination of the path of a projectile in a medium which resists proportionally to the square of the velocity, that being nearly the law of the resistance which the air opposes to bodies moving with great velocity. The resistance of fluids had been treated of by Newton in the second book of the Principia, and he had investigated a great number of curious and important propositions relative to its effects. He had considered some of the simpler laws of resistance, but of the case just mentioned he had given no solution, and, after approaching as near as possible to it on all sides, had withdrawn without making an attack. A problem so formidable was not likely to meet with many who, even in the more improved state at which the calculus had now arrived, could hope to overcome its difficulties. Whether Keill had flattered himself that he could resolve the problem, or had forgotten, that when a man proposes a question of defiance to another, he ought to be sure that he can answer it himself, may be doubted; but this is certain, that, without the necessary preparation, he boldly challenged Bernoulli to produce a solution.

Bernoulli resolved the question in a very short time, not only for a resistance proportional to the square, but to any power whatsoever of the velocity, and by the conditions which he affixed to the publication of his solution, took care to expose the weakness of his antagonist. He repeatedly offered to send his solution to a confidential person in London, providing Keill would do the same. Keill never made any reply to a proposal so fair, that there could only be one reason for declining it. Bernoulli, of course, exulted over him cruelly, breaking out in a torrent of vulgar abuse, and losing sight of every maxim of candour and good taste.

Such, then, were the circumstances under which the infinitesimal analysis,—the greatest discovery ever made in the mathematical sciences,—was ushered into the world. Every where, as it became known, it enlarged the views, roused the activity, and increased the power of the geometer, while it directed the warmest sentiments of his gratitude and admiration toward the great inventors. In one respect, only, its effects were different from those which one would have wished to see produced. It excited jealousy between two great men who ought to have been the friends of one another, and disturbed in both that philosophic tranquillity of mind, for the loss of which even glory itself is scarcely an adequate recompense.

In order to form a correct estimate of the magnitude and value of this discovery, it may be useful to look back at the steps by which the mathematical sciences had been prepared for it. When we attempt to trace those steps to their origin, we find the principle of the infinitesimal analysis making its first appearance in the method of Exhaustions, as exemplified in the writings of Euclid and Archimedes. These geometers observed, and, for what we know, were the first to observe, that the approach which a rectilineal figure may make to one that is curvilineal, by the increase of the number of its sides, the diminution of their magnitude, and a certain enlargement of the angles they contain, may be such that the properties of the former shall coincide so nearly with those of the latter, that no real difference can be supposed between them without involving a contradiction ; and it was in ascertaining the conditions of this approach, and in showing the contradiction to be unavoidable, that the method of Exhaustions consisted. The demonstrations were strictly geometrical, but they were often complicated, always indirect, and of course synthetical, so that they did not explain the means by which they had been discovered.

At the distance of more than two thousand years, Cavalieri advanced a step farther, and, by the sacrifice of some apparent, though of no real accuracy, explained, in the method of indivisibles, a principle which could easily be made to assume the more rigid form of Exhaustions. This was a very important discovery;—though the process was not analytical, the demonstrations were direct, and, when applied to the same subjects, led to the same conclusions which the ancient geometers had deduced; by an indirect proof also, such as those geometers had adopted, it could always be shown that an absurdity followed from supposing the results deduced from the method of indivisibles to be other than rigorously true.

The method of Cavalieri was improved and extended by a number of geometers of great genius who followed him; Torricelli, Roberval, Fermat, Huygens, Barrow, who all observed the great advantage that arose from applying the general theorems concerning variable quantity to the cases where the quantities approached to one another infinitely near, that is, nearer than within any assigned difference. There was, however, as yet, no calculus adapted to these researches, that is, no general method of reasoning by help of arbitrary symbols.

But we must go back a step, in point of time, if we would trace accurately the history of this last improvement. Descartes, as has been shown in the former part of this outline, made a great revolution in the mathematical sciences, by applying algebra to the geometry of curves; or, more generally, by applying it to express the relations of variable quantity. This added infinitely to the value of the algebraic analysis, and to the extent of its investigations. The same great mathematician had observed the advantage that would be gained in the geometry of curves, by considering the variable quantities in one state of an equation as differing infinitely little from the corresponding quantities in another state of the same equation. By means grounded on this he had attempted to draw tangents to curves, and to determine their curvature; but it is seldom the destination of Nature that a new discovery should be begun and perfected by the same individual; and, in these attempts, though Descartes did not entirely fail, he cannot be considered as having been successful. *

* See page 41.

At last came the two discoverers, Newton and Leibnitz, who completely lifted up the veil which their predecessors had been endeavouring to draw aside. They plainly saw, as Descartes indeed had done in part, that the infinitely small variations of the ordinate and abscissa are closely connected with many properties of the curve, which have but a very remote dependence on the ordinates and abscissæ themselves. Hence they inferred, that, to obtain an equation expressing the relations of these variations to one another, was to possess the most direct access to the knowledge of those properties. They observed also, that when an equation of this kind was deduced from the general equation, it admitted of being brought to great simplicity, and of being resolved much more readily than the other. In effect, it assumed the form of a simple equation; but, in order to make this deduction in the readiest and most distinct way, the introduction of new symbols, or of a new algorithm, was necessary, the invention of which could cost but little to the creative genius of the men of whom I now speak. They appear, as has been already shown, to have made their discoveries separately;—Newton first,—Leibnitz afterwards, at a considerable interval, yet the earliest, by several years, in communicating his discoveries to the world.

Thus, though there had been for ages a gradual

approach to the new analysis, there were in that progress some great and sudden advances which elevated those who made them to a much higher level than their predecessors. A great number of individuals co-operated in the work; but those who seem essential, and in the direct line of advancement, are Euclid, Cavalieri, Descartes, Newton, and Leibnitz. If any of the others had been wanting, the world would have been deprivėd of many valuable theorems, and many collateral improvements, but not of any general method essential to the completion of the infinitesimal analysis.

The views, however, of this analysis taken by the two inventors were not precisely the same. Leibnitz, considering the differences of the variable quantities as infinitely small, conceived that he might reject the higher powers of those differences without any sensible error; so that none of those powers but the first remained in the differential equation finally obtained. The rejection, however, of the higher powers of the differentials was liable to objection, for it had the appearance of being only an approximation, and did not come up to the perfect measure of geometrical precision. The analysis, thus constituted, necessarily divided itself into two problems;—the first is,—having given an equation involving two or more variable quantities, to find the equation expressing the relation of the differentials, or infinitely small variations of those quantities; the second is the converse of this;—having given an equation involving two or more variable quantities, and their differentials, to exterminate the differentials, and so to exhibit the variable quantities in a finite state. This last process is called *integration* in the language of the differential analysis, and the finite equation obtained is called the *integral* of the given differential equation.

Newton proceeded in some respects differently, and so as to preserve his calculus from the imputation of neglecting or throwing away any thing merely because it was small. Instead of the actual increments of the flowing or variable quantities, he introduced what he called the fluxions of those quantities,—meaning, by fluxions, quantities which had to one another the same ratio which the increments had in their ultimate or evanescent state. He did not reject quantities, therefore, merely because they were so small that he *might* do so without committing any sensible error, but because he *must* reject them, in order to commit no error whatsoever. Fluxions were, with him, nothing else than measures of the velocities with which variable or flowing quantities were supposed to be generated, and they might be of any magnitude, providing they were in the ratio of those velocities, or, which is the same, in the ratio of the nascent or evanescent increments.* The fluxions, therefore, and the flowing quantities or fluents of Newton correspond to the differentials and the sums or integrals of Leibnitz; and though the symbols which denote fluxions are different from those used to express differentials, they answer precisely the same purpose. The fluxionary and differential calculus may therefore be considered as two modifications of one general method, aptly distinguished by the name of the *infinitesimal analysis.*

By the introduction of this analysis, the domain of the mathematical sciences was incredibly enlarged in every direction. The great improvement which Descartes had made by the application of algebraic equations to define the nature of curve lines was now rendered much more efficient, and carried far beyond its original boundaries. From the equation of the curve the new analysis could deduce the properties of the tangents, and, what was much more difficult, could go back from the properties of the tangents to the equation of the curve. From the same equation it was able to determine the cur-

* "I consider mathematical quantities in this place not as consisting of small parts, but as described by a continued motion. Lines are described and thereby generated, not by the apposition of parts, but by the continued motion of points, superficies by the motion of lines," &c.—*Quadrature of Curves,* Introduction.

vature at every point; it could measure the length of any portion of the curve or the area corresponding to it. Nor was it only to algebraic curves that those applications of the calculus extended, but to curves transcendental and mechanical, as in the instances of the catenaria, the cycloid, the elastic curve, and many others. The same sort of research could be applied to curve surfaces described according to any given law, and also to the solids contained by them.

The problems which relate to the *maxima* and *minima*, or the greatest and least values of variable quantities, are among the most interesting in the mathematics; they are connected with the highest attainments of wisdom and the greatest exertions of power; and seem like so many immoveable columns erected in the infinity of space, to mark the eternal boundary which separates the regions of possibility and impossibility from one another. For the solution of these problems, a particular provision seemed to be made in the new geometry.

When any function becomes either the greatest or the least, it does so by the velocity of its increase or of its decrease ceasing entirely, or, in the language of algebra, becoming equal to nothing. But when the velocity with which the function varies becomes nothing, the fluxion which is proportional to that velocity must become nothing also. Therefore, it is only necessary to take the fluxion of the given function, and, by supposing it equal to nothing, an equation will be obtained in finite terms, (for the fluxion will entirely disappear,) expressing the relation of the quantities when the function assigned is the greatest or the least possible.

Another kind of maximum or minimum, abounding also in interesting problems, is more difficult by far than the preceding, and, when taken generally, seems to be only accessible to the new analysis. Such cases occur when the function of the variable quantities which is to be the greatest or the least is not given, but is itself the thing to be found ; as when it is proposed to determine the line by which a heavy body can descend in the least time from one point to another. Here the equation between the co-ordinates of the curve to be found is, of course, unknown, and the function of those co-ordinates which denotes the time of descent cannot, therefore, be algebraically expressed, so that its fluxion cannot be taken in the ordinary way, and thus put equal to nothing. The former rule, then, is not applicable in such cases, and it is by no means obvious in what manner this difficulty is to be overcome. The general problem exercised the ingenuity of both the Bernoullis, as it has since done of many other mathematicians of the greatest name. As there are in such problems always two conditions, according to the first of which, a certain property is to remain constant, or to belong to all the individuals of the species, and, according to the second, another property is to be the greatest or the least possible; and as, in some of the simplest of such questions,* the constant quantity is the circumference or perimeter of a certain curve, so problems of this kind have had the name of *Isoperimetrical* given them, a term which has thus come to denote one of the most curious and difficult subjects of mathematical investigation.

The new analysis, especially according to the view taken of it by Newton, is peculiarly adapted to physical researches, as the hypothesis of quantities being generated by continued motion, comes there to coincide exactly with the fact. The momentary increments or the fluxions represent so precisely the forces by which the changes in nature are produced, that this doctrine seemed created for the express purpose of penetrating into the interior of things, and taking direct cognizance of those animating powers which, by their subtility, not only elude the observation of sense, but the ordinary

* The most simple problem of the kind is strictly and literally *Isoperimetrical*, viz. of all curves having the same perimeter to find that which has the greatest area. Elementary geometry had pronounced this curve to be the circle long before there was any idea of an entire class of problems characterized by similar conditions. *Vide* Pappi Alexandrini Collect. Math. Lib. V. Prop. 2, &c.

methods of geometrical investigation. The infinitesimal analysis alone affords the means of measuring forces, when each acts separately, and instantaneously under conditions that can be accurately ascertained. In comparing the effects of continued action, the variety of time and circumstance, and the continuance of effects after their causes have ceased, introduce so much uncertainty, that nothing but vague and unsatisfactory conclusions can be deduced. The analysis of infinites goes directly to the point; it measures the intensity or instantaneous effort of the force, and, of course, removes all those causes of uncertainty which prevailed when the results of *continued* action could alone be estimated. It is not even by the effects produced in a short time, but by effects taken in their *nascent* or *evanescent* state, that the true proportion of causes must be ascertained.

Thus, though the astronomers had proved that the planets describe ellipses round the sun as the common focus, and that the line from the sun to each planet sweeps over areas proportional to the time; had not the geometer resolved the elliptic motion into its primary elements, and compared them in their state of evanescence, it would never have been discovered that these bodies gravitate to the sun with forces which are inversely as the square of their distances from the centre of that luminary. Thus, fortunately, the first discovery of

Newton was the instrument which was to conduct him safely through all the intricacies of his future investigations.

The calculus, as already remarked, necessarily divides itself into two branches; one which, from the variable quantities, finds the relation of their fluxions or differentials; another which, from the relation of these last, investigates the relation of the variable quantities themselves. The first of these problems is always possible, and, in general, easy to be resolved; the second is not always possible, and when possible, is often very difficult, but in various degrees, according to the manner in which the differentials and the variable quantities are combined with one another.

If the function, into which the differential stands multiplied, consist of a single term, or an aggregate of terms, in each of which the variable quantity is raised to a power expounded by a number positive, negative, or fractional, the integration can be effected with ease, either in algebraic or logarithmic terms; and the calculus had not been long known before this problem was completely resolved.

The second case of this first division is, when the given function is a fraction having a binomial or multinomial denominator, the terms of which contain any powers whatever of the variable magnitude, but without involving the radical sign. If the denominator contain only the simple power of the variable quantity, the integral is easily found by logarithms; if it be complex, it must be resolved either into simple or quadratic divisors, which, granting the solution of equations, is always possible, at least by approximation, and the given fraction is then found equal to an aggregate of simple fractions, having these divisors for their denominators, and of which the fluents can always be exhibited in algebraic terms, or in terms of logarithms and circular arches. This very general and important problem was resolved by J. Bernoulli as early as the year 1702.

The denominator is in this last case supposed rational; but if it be irrational, the integration requires other means to be employed. Here Leibnitz and Bernoulli both taught, how, by substitutions, as in *Diophantine* problems, the irrationality might be removed, and the integration of course reduced to the former case. Newton employed a different method, and, in his *Quadrature of Curves*, found the fluents, by comparing the given fluxion with the formulas immediately derived from the expression of circular or hyperbolic areas. The integrations of these irrational formulæ, whichever of the methods be employed, often admit of being effected with singular elegance and simplicity; but a general integration of all the formulæ of this kind, except by approximation, is not yet within the power of analysis.

The second general division of the problem of integration, viz. when the two variable quantities and their differentials are mixed together on each side of the equation, is a more difficult subject of inquiry than the preceding. It may indeed happen, that an equation, which at first presents itself under this aspect, can, by the common rules of algebra, have the quantities so separated, that on each side of the sign of equality there shall be but one variable quantity with its fluxion; and when this is done, the integration is reduced to one of the cases already enumerated.

When such separation cannot be made, the problem is among the most difficult which the infinitesimal analysis presents, at the same time that it is the key to a vast number of interesting questions both in the pure and the mixed mathematics. The two Bernoullis applied themselves strenuously to the elucidation of it; and to them we owe all the best and most accurate methods of resolving such questions which appeared in the early history of the calculus, and which laid the foundation of so many subsequent discoveries. This is a fact which cannot be contested; and it must be acknowledged also, that, on the same subject, the writings of the English mathematicians were then, as they continue to be at this day, extremely defective. Newton, though he had treated of this branch of the infinitesimal analysis with his usual ingenuity and depth, had done so only in his work on *Fluxions*, which did not see the light till several years after his death, when, in 1736, it appeared in Colson's translation. But that work, even had it come into the hands of the public in the author's lifetime, would not have remedied the defect of which I now speak. When the fluxionary equation could not be integrated by the simplest and most elementary rules, Newton had always recourse to approximations by infinite series, in the contrivance of which he indeed displayed great ingenuity and address. But an approximation, let it be ever so good, and converge ever so rapidly, is always inferior to an accurate and complete solution, if this last possess any tolerable degree of simplicity. The series which affords the approximation cannot converge always, or in all states of the variable quantity; and its utility, on that account, is so much limited, that it can hardly lead to any general result. Besides, it does not appear that these series can always be made to involve the arbitrary or indeterminate quantity, without which no fluent can be considered as complete. For these reasons, such approximations should never be resorted to till every expedient has been used to find an accurate solution. To this rule, however, Newton's method does not conform, but employs approximation in cases where the complete integral can be obtained. The tendency of that method, therefore, however great its merit in other respects, was to give a direction to research which was not always the best, and which, in many instances, made it fall entirely short of the object it ought to have attained. It is true, that many fluxionary equations cannot be integrated in any other way; but by having recourse to it indiscriminately, we overlook the cases in which the integral can be exactly assigned. Accordingly, Bernoulli, by following a different process, remarked entire classes of fluxionary or differential equations, that admitted of accurate integration. Thus he found, that differential equations, if homogeneous, * however complicated, may always have the variable quantities separated, so as to come under one of the simpler forms already enumerated. By the introduction, also, of exponential equations, which had been considered in England as of little use, he materially improved this branch of the calculus.

To all these branches of analysis we have still another to add of indefinite extent, arising out of the consideration of the fluxions or differentials of the higher orders, each of these orders being deduced from the preceding, just as first fluxions are from the variable quantities to which they belong.

---

* Homogeneous equations in the differential calculus are those in which the sum of the exponents of the variable quantities is the same in all the terms.

To understand this, conceive the successive values of the first fluxions of any variable quantity, to constitute a new series of variable quantities flowing with velocities, the measures of which form the fluxions of the second order, from which, in the same manner, are deduced fluxions of the third and of still higher orders. The general principles are the same as in the fluxions of the first order, but the difficulties of the calculus are greater, particularly in the integrations ; for to rise from second fluxions to the variable quantities themselves two integrations are necessary ; from third fluxions three, and so on.

The tract which first made known the new analysis was that of Leibnitz, published, as already remarked, in the first volume of the *Acta Eruditorum* for 1684, where it occupies no more than six pages,* and is the work of an author not yet become very familiar with the nature of his own invention. It was sufficient, however, to explain that invention to mathematicians ; but, nevertheless, some years elapsed before it drew much attention. The Bernoullis were the first who perceived its value, and made themselves masters of the principles and methods contained, or rather suggested, in it. Leibnitz published many other papers in the *Acta Eru-*

* Nova Methodus pro Maximis et Minimis, &c. Leibnitii Opera, Tom. III. p. 167.

ditorum and the journals of the times, full of original views and important hints, thrown out very briefly, and requiring the elucidations which his friends just mentioned were always so willing and so able to supply. The number of literary and scientific objects which divided the attention of the author himself was so great, that he had not time to bestow on the illustration and developement of the most important of his own discoveries, and the new analysis, for all that he has taught, would have been very little known, and very imperfectly unfolded, if the two excellent geometers just named had not come to his assistance. Their tracts were also, like his, scattered in the different periodic works of that time, and several years elapsed before any elementary treatise explained the general methods, and illustrated them by examples. The first book in which this was done, so far at least as concerned the differential or direct calculus, was the Analyse des Infiniment Petits of the Marquis de l'Hôpital, published in 1696, a work of great merit, which did much to diffuse the knowledge of the new analysis. It was well received at that time, and has maintained its character to the present day. The author, a man of genius, indefatigable and ardent in the pursuit of science, had enjoyed the *viva voce* instructions of John Bernoulli, on the subject of the new geometry, and therefore came forward with every possible advantage.

It was long after this before the works of the Bernoullis were collected together, those of James in two quarto volumes, and of John in four.* In the third of these last volumes is a tract of considerable length, with the title of Lectiones de Methodo Integralium, written in 1691 and 1692, for the use of M. de l'Hôpital, to whose book on the differential calculus it seems to have been intended as a sequel. It is a work of great merit; and affords a distinct view of many of the most general methods of integration, with their application to the most interesting problems; so that, though the earliest treatise on that subject, it remains at this day one of the best compends of the new analysis of which the mathematical world is in possession. Indeed, the whole of the volumes just referred to are highly interesting, as containing the original germs of the new analysis, and as being the work of men always inspired by genius, sometimes warmed by opposition, and generally animated by the success which accompanied their researches.

But we must now look at the original works of the earliest inventor. Newton, besides his letters published in the Commercium Epistolicum, is the author of three tracts on the new analysis that have all been occasionally mentioned. None of them, however, appeared nearly so soon as a great num-

* Those of James were published at Geneva in 1744; of John at Lausanne and Geneva in 1742.

ber of the pieces which have just been enumerated. The Quadrature of Curves, written as early as 1665 or 1666, did not appear till 1704; and though it be a treatise of great value, and containing very important and very general theorems concerning the quadrature of curves, it must be allowed, that it is not well adapted to make known the spirit and the views of the infinitesimal analysis. After a short introduction, which is indeed analytical, and which explains the idea of a fluxion with great brevity and clearness, the treatise sets out with proposing to find any number of curves that can be squared; and here the demonstrations become all synthetical, without any thing that may be properly called analytical investigation. By synthetical demonstrations I do not mean reasonings where the algebraic language is not used, but reasonings, whatever language be employed, where the solution of the proposed question is first laid down, and afterwards demonstrated to be true. Such is the method pursued throughout this work, and it is wonderful how many valuable conclusions concerning the areas of curves, and their reduction to the areas of the circle and hyperbola, are in that manner deduced. But though truths can be very well conveyed in the synthetical way, the methods of investigating truth are not communicated by it, nor the powers of invention directed to their proper objects. As an elementary treatise on the new analysis, the

Quadrature of Curves is therefore imperfect, and not calculated, without great study, to give to others any portion of the power which the author himself has exerted. The problem of finding fluents, though it be that on which the whole quadrature of curves depends, is entirely kept out of view, and never once proposed in the course of a work, which, at the same time, is full of the most elaborate and profound reasonings.

Newton had a great fondness for the synthetical method, which is apparent even in the most analytical of his works. In his Fluxions, when he is treating of the quadrature of curves, he says, "After the area of a curve has been found and constructed, we should consider about the demonstration of the construction, that, laying aside all algebraical calculation, as much as may be, the theorem may be adorned and made elegant, so as to become fit for public view." * This is followed by two or three examples, in which the rule here given is very happily illustrated. When the analysis of a problem requires, like the quadrature of curves, the use of the inverse method of fluxions, the reversion of that analysis, or the synthetical demonstration, must proceed by the direct method, and therefore may admit of more simplicity than the

* Newton's Fluxions, Colson's Translation, p. 116, § 107.

others, so as, in the language of the above passage, to be easily adorned and made elegant.

The book of Fluxions is, however, an excellent work, entering very deeply into the nature and spirit of the calculus,—illustrating its application by well chosen examples,—and only failing, as already said, by having recourse, for finding the fluents of fluxionary equations, too exclusively to the method of series, without treating of the cases in which exact solutions can be obtained.

Of the works that appeared in the early stages of the calculus, none is more entitled to notice than the Harmonia Mensurarum of Cotes. The idea of reducing the areas of curves to those of the circle and hyperbola, in those cases which did not admit of an accurate comparison with rectilineal spaces, had early occurred to Newton, and was very fully exemplified in his Quadrature of Curves. Cotes extended this method:—his work appeared in 1722, and gave the rules for finding the fluents of fractional expressions, whether rational or irrational, greatly generalized and highly improved by means of a property of the circle discovered by himself, and justly reckoned among the most remarkable propositions in geometry. It is singular that a work so profound, and so useful as the Harmonia Mensurarum, should never have acquired, even among the mathematicians of England, the popularity which it deserves; and that, on the

Continent, it should be very little known, even after the excellent commentary and additions of Bishop Walmsley. The reasons, perhaps, are, that, in many parts, the work is obscure; that it does not explain the analysis which must have led to the *formulæ* contained in the tables; and that it employs an unusual language and notation, which, though calculated to keep in view the analogy between circular and hyperbolic areas, or between the measures of angles and of ratios, do not so readily accommodate themselves to the business of calculation as those which are commonly in use. Demoivre, a very skilful and able mathematician, improved the method of Cotes; and explained many things in a manner much more clear and analytical than had hitherto been done. *

Another very original and profound writer of this period was Brook Taylor, who has already been often mentioned, and who, in his Method of Increments, published in 1715, added a new branch to the analysis of variable quantity. According to this method, quantities are supposed to change, not by infinitely small, but by finite increments, or such as may be of any magnitude whatever. There are here, therefore, as in the case of

---

* Demoivre, Miscellanea Analytica. See also the work of an anonymous author, Epistola ad Amicum de Cotesii Inventis.

fluxions or differentials, two general questions: A function of a variable quantity being given, to find the expression for the finite increment of that function, the increment of the variable quantity itself being a finite magnitude. This corresponds to the direct method of fluxions; the other question corresponds to the inverse, viz. A function being given containing variable quantities, and their increments any how combined, to find the function from which it is derived. The author has considered both these problems, and in the solution of the second, particularly, has displayed much address. He has also made many ingenious applications of this calculus both to geometrical and physical questions, and, above all, to the summation of series, a problem for the solution of which it is peculiarly adapted.

Taylor, however, was more remarkable for the ingenuity and depth, than for the perspicuity of his writings; even a treatise on Perspective, of which he is the author, though in other respects excellent, has always been complained of as obscure; and it is no wonder if, on a new subject, and one belonging to the higher geometry, his writings should be still more exposed to that reproach. This fault was removed, and the whole theory explained with great clearness, by M. Nicol, of the Academy of Sciences of Paris, in a series of Mémoires from the year 1717 to 1727.

A single analytical formula in the *Method of Increments* has conferred a celebrity on its author, which the most voluminous works have not often been able to bestow. It is known by the name of Taylor's Theorem, and expresses the value of any function of a variable quantity in terms of the successive orders of increments, whether finite or infinitely small. If any one proposition can be said to comprehend in it a whole science it is this: for from it almost every truth and every method of the new analysis may be deduced. It is difficult to say, whether the theorem does most credit to the genius of the author, or the power of the language which is capable of concentrating such a vast body of knowledge in a single expression. Without an acquaintance with algebra, it is impossible, I believe, to conceive the manner in which this effect is produced.

By means of its own intrinsic merit, and the advantageous display of it made in the works now enumerated, the new analysis, long before the expiration of the period of which I am here treating, was firmly established all over Europe. It did not, however, exist everywhere in the same condition, nor under the same form; with the British and Continental mathematicians, it was referred to different origins; it was in different states of advancement; the notation and some of the fundamental ideas were also different. The authors communicated little with one another, except in the way of defiance or reproach'; and, from the angry or polemical tone which their speculations often assumed, one could hardly suppose, that they were pursuing science in one of its most abstract and incorporeal forms.

Though the algorithm employed, and the books consulted on the new analysis, were different, the mathematicians of Britain and of the Continent had kept pace very nearly with one another during the period now treated of, except in one branch, the integration of differential or of fluxional equations. In this, our countrymen had fallen considerably behind, as has been already explained; and the distance between them and their brethren on the Continent continued to increase, just in proportion to the number and importance of the questions, physical and mathematical, which were found to depend on these integrations. The habit of studying only our own authors on these subjects, produced at first by our admiration of Newton and our dislike to his rivals, and increased by a circumstance very insignificant in itself, the diversity of notation, prevented us from partaking in the pursuits of our neighbours; and cut us off in a great measure from the vast field in which the genius of France, of Germany, and Italy, was exercised with so much activity and success. Other causes may have united in the production of an effect, which the mathematicians of this country have had much reason to regret; but the evil had its origin in the spirit of jealousy and opposition, which arose from the controversies that have just passed under our review. The habits so produced continued long after the spirit itself had subsided.

It must not be supposed, that so great a revolution in science, as that which was made by the introduction of the new analysis, could be brought about entirely without opposition, as in every society there are some who think themselves interested to maintain things in the condition wherein they have found them. The considerations are indeed sufficiently obvious, which, in the moral and political world, tend to produce this effect, and to give a stability to human institutions, often so little proportionate to their real value or to their general utility. Even in matters purely intellectual, and in which the abstract truths of arithmetic and geometry seem alone concerned, the prejudices, the selfishness, or vanity of those who pursue them, not unfrequently combine to resist improvement, and often engage no inconsiderable degree of talent in drawing back instead of pushing forward the machine of science. The introduction of methods entirely new must often change the relative place of the men engaged in scientific pursuits; and must oblige many, after descending from the stations they formerly occupied, to take a lower position in the scale of intellectual advancement. The enmity of such men, if they be not animated by a spirit of real candour and the love of truth, is likely to be directed against methods, by which their vanity is mortified, and their importance lessened. Though such changes as this must have everywhere accompanied the ascendancy acquired by the calculus, for the credit of mathematicians it must be observed, that no one of any considerable eminence has had the misfortune to enrol his name among the adversaries of the new science; and that Huygens, the most distinguished and most profound of the older mathematicians then living, was one of the most forward to acknowledge the excellence of that science, and to make himself master of its rules, and of their application.

Nevertheless, certain adversaries arose successively in Germany, France, and England, the countries in which the new methods first became known.

Nieuwentyt, an author commendable as a naturalist, and as a writer on morals, but a very superficial geometer, aimed the first blow at the Differential Calculus. He objected to the explanation of Leibnitz, and to the notion of quantities infinitely small.* It seemed as if he were unwilling to be-

* He published *Analysis Infinitorum* at Amsterdam, in 1695; and another tract, *Considerationes circa Calculi Differentialis Principia,* in the year following. This last was answered by Herman.

lieve in the reality of objects smaller than those discovered by his own microscope, and were jealous of any one who should come nearer to the limit of extension than he himself had done. Leibnitz thought his objections not undeserving of a reply; but the reply was not altogether satisfactory. A second was given with better success; and afterwards Herman and Bernoulli each severally defeated an adversary, who was but very ill able to contend with either of them.

Soon after this, the calculus had to sustain an attack from two French academicians, which drew more attention than that of the Dutch naturalist. One of these, Rolle, was a mathematician of no inconsiderable acquirement, but whose chief gratification consisted in finding out faults in the works of others. He founded his objections to the differential calculus, not on the score of principles or of general methods, but on certain cases which he had sought out with great industry, in which those methods seemed to him to lead to false and contradictory conclusions. On examination, however, it turned out, that in every one of those instances the error was entirely his own; that he had misapplied the rules, and that his eagerness to discover faults had led him to commit them. His errors were detected and pointed out with demonstrative evidence by Varignon, Saurin, and some others, who were among the first to perceive the excellence and to defend the solidity of the new geometry. These disputes were of consequence enough to occupy the attention of the Academy of Sciences during a great part of the year 1701.

The Abbé Gallois joined with Rolle in his hostility to the calculus, and though he added very little to the force of the attack, he kept the field after the other had retired from the combat. Fontenelle, in his *éloge* on the Abbé, has given an elegant turn to the apology he makes for him.—" His taste for antiquity made him suspicious of the geometry of infinites. He was, in general, no friend to any thing that was new, and was always prepared with a kind of *Ostracism* to put down whatever appeared too conspicuous for a free state like that of letters. The geometry of infinites had both these faults, and particularly the latter."

After all these disputes were quieted in France, and the new analysis appeared completely victorious, it had an attack to sustain in England from a more formidable quarter. Berkeley Bishop of Cloyne was a man of first-rate talents, distinguished as a metaphysician, a philosopher, and a divine. His geometrical knowledge, however, which, for an attack on the method of fluxions, was more essential than all his other accomplishments, seems to have been little more than elementary. The motive which induced him to enter on discussions so remotely connected with his usual pursuits has been variously represented; but, whatever it was, it gave rise to the Analyst, in which the author professes to demonstrate, that the new analysis is inaccurate in its principles, and that, if it ever lead to true conclusions, it is from an accidental compensation of errors that cannot be supposed always to take place. The argument is ingeniously and plausibly conducted, and the author sometimes attempts ridicule with better success than could be expected from the subject; thus, when he calls ultimate ratios the *ghosts of departed quantitics*, it is not easy to conceive a witty saying more happily fastened on a mere mathematical abstraction.

The Analyst was answered by Jurin, under the signature of *Philalethes;* and to this Berkeley replied in a tract entitled A Defence of Freethinking in Mathematics. Replies were again made to this, so that the argument assumed the form of a regular controversy; in which, though the defenders of the calculus had the advantage, it must be acknowledged that they did not always argue the matter quite fairly, nor exactly meet the reasoning of their adversary. The true answer to Berkeley was, that what he conceived to be an accidental compensation of errors was not at all accidental, but that the two sets of quantities that seemed to him neglected in the reasoning were in all cases necessarily equal, and an exact balance for one another. The Newtonian idea of a fluxion contained in it this truth, and so it was argued by Jurin and others, but not in a manner so logical and satisfactory as might have been expected. Perhaps it is not too much to assert, that this was not completely done till La Grange's Theory of Functions appeared. Thus, if the author of the Analyst has had the misfortune to enrol his name on the side of error, he has also had the credit of proposing difficulties of which the complete solution is only to be derived from the highest improvements of the calculus.

This controversy made some noise in England, but I do not think that it ever drew much attention on the Continent. The Analyst, I imagine, notwithstanding its acuteness, never crossed the Channel. Montucla evidently knows it only by report, and seems as little acquainted with the work as with its author, of whom he speaks very slightly, and supposes he has sufficiently described him by saying, that he has written a book against the existence of matter, and another in praise of tar-water. But it is less from the opinions which men support than from the manner in which they support them, that their talents are to be estimated. If we judge by this criterion, we shall pronounce Berkeley to be a man of genius, whether he be employed in attacking the infinitesimal analysis, in disproving the existence of the external world, or in celebrating the virtues of tar-water. *

---

* Though Berkeley reasons very plausibly, and with considerable address, he hurts his cause by the comparison so often introduced between the mysteries of religion and what he accounts the mysteries of the new geometry. From this it is natural to infer, that the author is avenging the cause of religion on the infidel mathematician to whom his treatise is addressed, and an argument that is suspected to have any other object than that at which it is directly aimed, must always lose somewhat of its weight.

The dispute here mentioned did not take place till about the year 1734; so that I have here treated of it by anticipation, being unwilling to resume the subject of controversies which, though perhaps useful at first for the purpose of securing the foundations of science, are long since set to rest, and never likely to be revived.

SECTION II.

MECHANICS, GENERAL PHYSICS, &c.

THE discoveries of Galileo, Descartes, and other mathematicians of the seventeenth century, had made known some of the most general and important laws which regulate the phenomena of moving bodies. The inertia, or the tendency of body, when left to itself, to preserve unchanged its condition either of motion or of rest; the effect of an impulse communicated to a body, or of two simultaneous impulses, had been carefully examined, and had led to the discovery of the composition of motion. The law of equilibrium, not in the lever alone, but in all the mechanical powers, had been determined, and the equality of action to reaction, or of the motion lost to the motion acquired, had not only been established by reasoning, but confirmed by experiment. The fuller elucidation and farther extension of these principles were reserved for the period now treated of.

The developement of truth is often so gradual, that it is impossible to assign the time when certain principles have been first introduced into science. Thus, the principle of *Virtual Veloci-*

ties, as it is termed, which is now recognized as regulating the equilibrium of all machines whatsoever, was perceived to hold in particular cases long before its full extent, or its perfect universality, was understood. Galileo made a great step toward the establishment of this principle when he generalized the property of the lever, and showed, that an equilibrium takes place whenever the sums of the opposite *momenta* are equal, meaning by momentum the product of the force into the velocity of the point at which it is applied. This was carried farther by Wallis, who appears to have been the first writer who, in his Mechanica, published in 1669, founded an entire system of statics on the principle of Galileo, or the equality of the opposite momenta. The proposition, however, was first enunciated in its full generality, and with perfect precision, * by John Bernoulli, in a letter

* The principle of Virtual Velocities may be thus enunciated:—If a system of bodies be in a state of equilibrium, in consequence of the action of any forces whatever, on certain points in the system; then were the equilibrium to be for a moment destroyed, the small space moved over by each of these points will express the virtual velocity of the power applied to it, and, if each force be multiplied into its virtual velocity, the sum of all the products where the velocities are in the same direction, will be equal to the sum of all those in which they are in the opposite.

The distinction between actual and virtual velocities was

to Varignon, so late as the year 1717. Varignon inserted this letter at the end of the second edition of his Projet d'une Nouvelle Mécanique, which was not published till 1725. The first edition of the same book appeared in 1687, and had the merit of deriving the whole theory of the equilibrium of the mechanical powers, from the single principle of the composition of forces. At first sight there appear in mechanics two independent principles of equilibrium, that of the lever, or of equal and opposite momenta, and that of the composition of forces. To show that these coincide, and that the one may be deduced from the other, is, therefore, doing a service to science, and this the ingenious author just named accomplished by help of a property of the parallelogram, which he seems to have been the first who demonstrated.

The Principia Mathematica of Newton, published also in 1687, marks a great era in the history of human knowledge, and had the merit of effecting an almost entire revolution in mechanics, by giving new powers and a new direction to its researches. In that work the composition of forces was treated independently of the composition of motion, and the equilibrium of the lever was de-

---

first made by Bernoulli, and is very essential to thinking as well as to speaking with accuracy on the nature of equilibriums.

duced from the former, as well as in the treatise already mentioned. From the equality of action and re-action it was also inferred, that the state of the centre of gravity of any system of bodies, is not changed by the action of those bodies on one another. This is a great proposition in the mechanics of the universe, and is one of the steps by which that science ascends from the earth to the heavens; for it proves that the quantity of motion existing in nature, when estimated in any one given direction, continues always of the same amount.

But the new applications of mechanical reasoning,—the reduction of questions concerning force and motion to questions of pure geometry,—and the mensuration of mechanical action by its nascent effects,—are what constitute the great glory of the Principia, considered as a treatise on the theory of motion. A transition was there made from the consideration of forces acting at stated intervals, to that of forces acting continually,—and from forces constant in quantity and direction to those that converge to a point, and vary as any function of the distance from that point; the proportionality of the areas described about the centre of force, to the times of their description; the equality of the velocities generated in descending through the same distance by whatever route;

the relation between the squares of the velocities produced or extinguished, and the sum of the accelerating or retarding forces, computed with a reference, not to the time during which, but to the distance over which they have acted. These are a few of the mechanical and dynamical discoveries contained in the same immortal work; a fuller account of which belongs to the history of physical astronomy.

The end of the seventeenth and the beginning of the eighteenth centuries were rendered illustrious, as we have already seen, by the mathematical discoveries of two of the greatest men who have ever enlightened the world. A slight sketch of the improvements which the theory of mechanics owes to Newton has been just given; those which it owes to Leibnitz, though not equally important, nor equally numerous, are far too conspicuous to be passed over in silence. So far as concerns general principles they are reduced to three,—the argument of the sufficient reason,—the law of continuity,—and the measurement of the force of moving bodies by the square of their velocities; which last, being a proposition that is true or false according to the light in which it is viewed, I have supposed it placed in that which is most favourable.

With regard to the first of these,—*the principle of the sufficient reason*,—according to which nothing exists in any state without a reason determining it to be in that state rather than in any other, —though it be true that this proposition was first distinctly and generally announced by the philosopher just named, yet is it certain that, long before his time, it had been employed by others in laying the foundations of science. Archimedes and Galileo had both made use of it, and, perhaps, there never was any attempt to place the elementary truths of science on a solid foundation in which this principle had not been employed. We have an example of its application in the proof usually given, that a body in motion cannot change the direction of its motion, abstraction being made from all other bodies, and from all external action; for it is evident, that no reason exists to determine the change of motion to be in one direction more than another, and we therefore conclude, that no such change can possibly take place. Many other instances might be produced where the same principle appears as an axiom of the clearest and most undeniable evidence. Wherever, indeed, we can pronounce, with certainty, that the conditions which determine two different things, whether magnitudes or events, are in two cases precisely the same, it cannot be doubted that these events or magnitudes are in all respects identical.

However sound this principle may be in itself, the use which Leibnitz sometimes made of it has tended to bring it into discredit. He argued, for example, that of the particles of matter no two can possess exactly the same properties, or can perfectly resemble one another, otherwise the Supreme Being could have no reason for employing one of them in a particular position more than another, so that both must necessarily be rejected. To argue thus, is to suppose that we completely understand the manner in which motives act on the mind of the Divinity, * a postulate that seems but ill suited to the limited sphere of the human understanding. But, if Leibnitz has misapplied his own principle, and extended its authority too far, this affords no ground for rejecting it when we are studying the ordinary course of nature, and arguing about the subjects of experiment and observation. In fact, therefore, the sciences which aspire to place their foundation on the solid basis of necessary truth, are much indebted to Leibnitz for the introduction of this principle into philosophy.

* The argument of Leibnitz seems evidently inconclusive. For, though there were two similar and equal atoms, yet as they could not co-exist in the same space, they would not, as far as position is concerned, bear the same relation to the particles that surrounded them ; there might exist, therefore, considering them as part of the materials to be employed in the construction of the universe, very good reasons for assigning different situations to each.

Another principle of great use in investigating the laws of motion, and of change in general, was brought into view by the same author,—*the law of Continuity*,—according to which, nothing passes from one state to another without passing through all the intermediate states. Leibnitz considers himself as the first who made known this law; but it is fair to remark, that, in as much as motion is concerned, it was distinctly laid down by Galileo,* and ascribed by him to Plato. But, though Leibnitz was not the first to discover the law of continuity, he was the first who regarded it as a principle in philosophy, and used it for trying the consistency of theories, or of supposed laws of nature, and the agreement of their parts with one another. It was in this way that he detected the error of Descartes's conclusions concerning the collision of bodies, showing, that though one case of collision must necessarily graduate into another, the conclusions of that philosopher did by no means pass from one to another by such gradual transition. Indeed, for the purpose of such detections, the knowledge of this law is extremely useful; and I believe few have been much occupied in the investigations either of the pure or mixed mathematics, who have not often been glad

* *Opere di Galileo,* Tom. III. p. 150, and Tom. II. p. 32. Ediz. di Padova, 1744.

to try their own conclusions by the test which it furnishes.

Leibnitz considered this principle as known *a priori*, because if any *saltus* were to take place, that is, if any change were to happen without the intervention of time, the thing changed must be in two different conditions at the same individual instant, which is obviously impossible. Whether this reasoning be quite satisfactory or not, the conformity of the law to the facts generally observed, cannot but entitle it to great authority in judging of the explanations and theories of natural phenomena.

It was the usual error, however, of Leibnitz and his followers, to push the metaphysical principles of science into extreme cases, where they lead to conclusions to which it was hardly possible to assent. The Academy of Sciences at Paris having proposed as a prize question, the Investigation of the Laws of the Communication of Motion, * John Bernoulli presented an Essay on the subject, very ingenious and profound, in which, however, he denied the existence of hard bodies, because, in the collision of such bodies, a finite change of motion must take place in an instant, an event which, on the principle just explained, he maintained to be impossible. Though the Essay was admired, this

* In 1724.

conclusion was objected to, and D'Alembert, in his *éloge* on the author, remarks, that, even in the collision of elastic bodies, it is difficult to conceive how, among the parts which first come into contact, a sudden change, or a change *per saltum*, can be avoided. Indeed, it can only be avoided by supposing that there is no real contact, and that bodies begin to act upon one another when their surfaces, or what seems to be their surfaces, are yet at a distance.

Maclaurin and some others are disposed, on account of the argument of Bernoulli, to reject the law of continuity altogether. This, however, I cannot help thinking, is to deprive ourselves of an auxiliary that, under certain restrictions, may be very useful in our researches, and is often so, even to those who profess to reject its assistance. It is admitted that the law of continuity generally leads right, and if it sometimes lead wrong, the true business of philosophy is to define when it may be trusted to as a safe guide, and what, on the other hand, are the circumstances which render its indications uncertain.

The discourse of Bernoulli, just referred to, brought another new conclusion into the field, and began a controversy among the mathematicians of Europe, which lasted for many years. It was a new thing to see geometers contending about the truths of their own science, and opposing one demonstration to another. The spectacle must have given pain to the true philosopher, but may have afforded consolation to many who had looked with envy on the certainty and quiet prevailing in a region from which they found themselves excluded.

Descartes had estimated the force of a moving body by the quantity of its motion, or by the product of its velocity into its mass. The mathematicians and philosophers who followed him did the same, and the product of these quantities was the measure of force universally adopted. No one, indeed, had ever thought of questioning the conformity of this measure to the phenomena of nature, when, in 1686, Leibnitz announced in the *Leipsic Journal*, the *demonstration of a great error committed by Descartes and others, in estimating the force of moving bodies*. In this paper, the author endeavoured to show, that the force of a moving body is not proportional to its velocity simply, but to the square of its velocity, and he supported this new doctrine by very plausible reasoning. A body, he says, projected upward against gravity, with a double velocity, ascends to four times the height; with the triple velocity, to nine times the height, and so on; the height ascended to being always as the square of the velocity. But the height ascended to is the effect, and is the natural measure of the force, therefore the force of a moving body is as the square of its velocity. Such was the first reasoning of Leibnitz on this subject,—simple, and apparently conclusive; nor should it be forgotten that, during the long period to which the dispute was lengthened out, and notwithstanding the various shapes which it assumed, the reasonings on his side were nothing more than this original argument, changed in its form, or rendered more complex by the combination of new circumstances, so as to be more bewildering to the imagination, and more difficult either to apprehend or to refute.*

John Bernoulli was at first of a different opinion from his friend and master, but came at length to adopt the same, which, however, appears to have gone no farther till the discourse was submitted to the Academy of Sciences, as has been already mentioned. The mathematical world could not look with indifference on a question which seemed to affect the vitals of mechanical science, and soon separated into two parties, in the arrangement of which, however, the effects of national predilection might easily be discovered. Germany, Holland, and Italy, declared for the *vis viva*; England stood firm for the old doctrine; and France was divided

* To mere pressure, Leibnitz gave the name of *vis mortua*, and to the force of moving bodies the name of *vis viva*. The former he admitted to be proportional to the simple power of the *virtual* velocity, and the second he held to be proportional to the square of the *actual* velocity.

between the two opinions. No controversy, perhaps, was ever carried on by more illustrious disputants; Maclaurin, Stirling, Desaguliers, Jurin, Clarke, Mairan, were all engaged on the one side, and on the opposite were Bernoulli, Herman, Poleni, S'Gravesende, Muschenbroek; and it was not till long after the period to which this part of the Dissertation is confined, that the debate could be said to be brought to a conclusion. That I may not, however, be obliged to break off a subject of which the parts are closely connected together, I shall take the liberty of transgressing the limits which the consideration of time would prescribe, and of now stating, as far as my plan admits of it, all that respects this celebrated controversy.

A singular circumstance may be remarked in the whole of the dispute. The two parties who adopted such different measures of force, when any mechanical problem was proposed concerning the action of bodies, whether at rest or in motion, resolved it in the same manner, and arrived exactly at the same conclusions. It was therefore evident, that, however much their language and words were opposed, their ideas or opinions exactly agreed. In reality, the two parties were not at issue on the question; their positions, though seemingly opposite, were not contrary to one another; and after debating for nearly thirty years, they found out this to be the truth. That the first men in the scientific world should have disputed so long with one another, without discovering that their opposition was only in words, and that this should have happened, not in any of the obscure and tortuous tracts through which the human mind must grope its way in anxiety and doubt, but in one of the clearest and straightest roads, where it used to be guided by the light of demonstration, is one of the most singular facts in the history of human knowledge.

The degree of acrimony and illiberality which were sometimes mixed in this controversy was not very creditable to the disputants, and proved how much more men take an interest in opinions as being their own, than as being simply in themselves either true or false. The dispute, as conducted by S'Gravesende and Clarke, took this turn, especially on the part of the latter, who, in the schools of theology having sharpened both his temper and his wit, accompanied his reasonings with an insolence and irritability peculiarly ill suited to a discussion about matter and motion. His paper on this subject, in the Philosophical Transactions, * contains many just and acute remarks, accompanied with the most unfair representation of the argument of his antagonists, as if the doctrine of the *vis viva*

* Vol. XXXV. (1728,) p. 381. Hutton's Abridgment, Vol. VII. p. 219.

were a matter of as palpable absurdity as the denial of one of the axioms of geometry.* Now, the truth is, that the argument in favour of living forces is not at all liable to this reproach. One of the effects produced by a moving body is proportional to the square of the velocity, while another is proportional to the velocity simply; and, according to which of these ways the force itself is to be measured, may involve the propriety or impropriety of mathematical language, but cannot be charged with absurdity or contradiction. Absurdity, indeed, was a reproach that neither side had any right to cast on the other.

A dissertation of Mairan, on the force of moving

---

* In all the arguments for the *vis viva,* this learned metaphysician saw nothing but a conspiracy formed against the Newtonian philosophy. "An extraordinary instance," says he, "of the maintenance of the most palpable absurdity we have had in late years of very eminent mathematicians, Leibnitz, Bernoulli, Herman, Gravesende, who, in order to raise a dust of opposition against the Newtonian philosophy, some years back insisted with great eagerness on a principle which subverts all science, and which easily may be made appear, even to an ordinary capacity, to be contrary to the necessary and essential nature of things." This passage may serve as a proof of the spirit which prevailed among the philosophers of that time, making them ascribe such illiberal views to one another, and distorting so entirely both their own reasoning and those of their adversaries. The spirit awakened by the discovery of fluxions had not yet subsided.

bodies, in the Mémoires of the Academy of Sciences for 1728, is one of those in which the common measure of force is most ably supported. Nevertheless, for a long time after this, the opinions on that subject in France continued still to be divided. In the list of the disputants we should hardly expect to find a lady included, if we did not know that the name of Madame de Chastellet, along with those of Hypatia and Agnesi, was honourably enrolled in the annals of mathematical learning. Her writings on this subject are full of ingenuity, though, from the fluctuation * of her opinions, it seems as if she had not yet entirely exchanged the caprice of fashion for the austerity of science. About the same time Voltaire engaged in the argu-

---

* Mad. de Chastellet, in a Dissertation on Fire, published in 1740, took the side of Mairan, and bestowed great praise on his discourse on the force of moving bodies. Having, however, afterwards become a convert to the philosophy of Leibnitz, she espoused the cause of the *Vis Viva*, and wrote against Mairan. At this time too she drew up a compend of the Leibnitian philosophy for the use of her son, which displays ingenuity and acuteness, and is certainly such a present as very few mothers have ever been in a condition to make to their children. Soon afterwards the same lady, having become a Newtonian, returned to her former opinion about the force of moving bodies, and in the end, gave to her countrymen an excellent translation of the Principia of Newton, with a commentary on a part of it, far superior to any other that has yet appeared.

ment, and in a Mémoire, * presented to the Academy of Sciences in 1741, contended that the dispute was entirely about words. His reasoning is on the whole sound, and the suffrage of one who united the character of a wit, a poet, and a philosopher, must be of great importance in a country where the despotism of fashion extends even to philosophical opinion.

The controversy was now drawing to a conclusion, † and in effect may be said to have been terminated by the publication of D'Alembert's Dynamique in 1743. I am not certain, however, that all the disputants acquiesced in this decision, at least till some years later. Dr Reid, in an essay On Quantity, in the Philosophical Transactions for 1748, has treated of this controversy, and remarked, that it had been dropt rather than concluded. In this I confess I differ from the learned author. The controversy seemed fairly ended, the arguments exhausted, and the conclusion established, that the propositions maintained by both

* Doutes sur la Mesure des Forces Motrices ; Œuvres de Voltaire, Tom. XXXIX, p. 91. 8vo edit. 1785.

† Two very valuable papers that appeared at this late period of the dispute are found in the Philosophical Transactions ; one by Desaguliers, in 1733, full of excellent remarks and valuable experiments ; another by Jurin, in 1745, containing a very full state of the whole controversy.

sides were true, and were not opposed to one another. Though the mathematical sciences cannot boast of never having had any debates, they can say that those that have arisen have always been brought to a satisfactory termination.

The observations with which I am to conclude the present sketch, are not precisely the same with those of the French philosopher, though they rest nearly on the same foundation.

As the effects of moving bodies, or the changes they produce, may vary considerably with accidental circumstances, we must, in order to measure their force, have recourse to effects which are uniform, and not under the influence of variable causes. First, we may measure the force of one moving body by its effect upon another moving body; and here there is no room for dispute, nor any doubt that the forces of such bodies are as the quantities of matter multiplied into the simple power of the velocities, because the forces of bodies in which these products are equal, are well known, if opposed, to destroy one another. Thus one effect of moving bodies affords a measure of their force, which does not vary as the *square;* but as the *simple power* of the velocity.

There is also another condition of moving bodies which may be expected to afford a simple and general measure of their force. When a moving body is opposed by pressure, by a *vis mortua*, or a resistance like that of gravity, the quantity of such resistance required to extinguish the motion, and reduce the body to rest, must serve to measure the force of that body. It is a force which, by repeated impulses, has annihilated another, and these impulses, when properly collected into one sum, must evidently be equal to the force which they have extinguished. It happens, however, that there are two ways of computing the amount of these retarding forces, which lead to different results, both of them just, and neither of them to be assumed to the exclusion of the other.

Suppose the body, the force of which is to be measured, to be projected perpendicularly upward with any velocity, then, if we would compute the quantity of the force of gravity which is employed in reducing it to rest, we may either inquire into the retardation which that force produces during a given time, or while the body is moving over a given space. In other words, we may either inquire how long the motion will continue, or how far it will carry the body before it be entirely exhausted. If the length of the time that the uniform resistance must act before it reduce the body to rest be taken for the effect, and consequently for the measure of the force of the body, that force must be proportional to the velocity, for to this the time is confessedly proportional. If, on the other hand, the length of the line which the moving body describes, while subjected to this uniform resistance, be taken for the effect and the measure of the force, the force must be as the square of the velocity, because to that quantity the line in question is known to be proportional. Here, therefore, are two results, or two values of the same thing, the force of a moving body, which are quite different from one another; an inconsistency which evidently arises from this, that the thing denoted by the term *force,* is too vague and indefinite to be capable of measurement, unless some farther condition be annexed. This condition is no other than a specification of the work to be performed, or of the effect to be produced by the action of the moving body. Thus, when to the question concerning the force of the moving body, you add that it is to be employed in putting in motion another body, which is itself free to move, no doubt remains that the force is as the velocity multiplied into the quantity of matter. So also, if the force of the moving body is to be opposed by a resistance like that of gravity, the length of time that the motion may continue is one of its measurable effects, and that effect is, like the former, proportional to the velocity. There is a third effect to be considered, and one which always occurs in such an experiment as the last,—the height to which the moving body will ascend. This limitation gives to the force a definite character, and it is now measured by the square of the velocity. In fact, therefore, it is not a precise question to ask, What is the measure of the force of a moving body? You must, in addition, say, How is the moving body to be employed, or in which of its different capacities is it that you would measure its effect? In this state of the question there is no ambiguity, nor any answer to be given but one. Hence it was that the mathematicians and philosophers who differed so much about the general question of the force of moving bodies, never differed about the particular applications of that force. It was because the condition necessary for limiting the vagueness and ambiguity of the *data*, in all such cases, was fully supplied.

In the argument, therefore, so strenuously maintained on the force of moving bodies, both sides were partly in the right and both partly in the wrong. Each produced a measure of force which was just in certain circumstances, and thus far had truth on his side: but each argued that his was the only true measure, so that all others ought to be rejected; and here each of them was in error. Hence, also, it is not an accurate account of the controversy to say that it was about words merely; the disputants did indeed misunderstand one another, but their error lay in ascribing generality to propositions that were true only in particular cases, to which, indeed, the ambiguity and vagueness of the word *force* materially contributed. It does not appear, however, that any good would now accrue from changing the language of dynamics. If, as has been already said, to the question, How are we to measure the force of a moving body? be added the nature of the effect which is to be produced, all ambiguity will be avoided.

It is, I think, only farther necessary to observe, that, when the resistance opposed to the moving body is not uniform but variable, according to any law, it is not simply either the time or the space which is proportional to the velocity or to the square of the velocity, but functions of those quantities. These functions are obtained from the integration of certain fluxionary expressions, in which the measures above described are applied, the resistance being regarded as uniform for an infinitely small portion of the time or of the space.

Many years after the period I am now treating of, the controversy about the *vis viva* seemed to revive in England, on the occasion of an Essay on *Mechanical Force*, by the late Mr Smeaton, an able engineer, who, to great practical skill, and much experience, added no inconsiderable knowledge of the mathematics.

The reality of the *vis viva*, then, under certain conditions, is to be considered as a matter completely established. Another inquiry concerning the nature of this force, which also gave rise to considerable debate, was, whether, in the communication of motion, and in the various changes through which moving bodies pass, the quantity of the *vis viva* remains always the same? It had been observed, in the collision of elastic bodies, that the *vis viva*, or the sum made up by multiplying each body into the square of its velocity, and adding the products together, was the same after collision that it was before it, and it was concluded with some precipitation, by those who espoused the Leibnitian theory, that a similar result always took place in the real phenomena of nature. Other instances were cited; and it was observed, that a particular view of this principle which presented itself to Huygens, had enabled him to find the centre of oscillation of a compound pendulum, at a time when the state of mechanical science was scarcely prepared for so difficult an investigation. The proposition, however, is true only when all the changes are gradual, and rigorously subjected to the law of continuity. Thus, in the collision of bodies imperfectly elastic, (a case which continually occurs in nature,) the force which, during the recoil, accelerates the separation of the bodies, does not restore to them the whole velocity they had lost, and the *vis viva*, after the collision, is always less than it was before it. The cases in which the whole amount of the *vis viva* is rigorously preserved, may always be brought under the thirty-ninth proposition of the first book of the Principia, where the principle of this theory is placed on its true foundation.

So far as General Principles are concerned, the preceding are the chief mechanical improvements which belong to the period so honourably distinguished by the names of Newton and Leibnitz. The application of these principles to the solution of particular problems would afford materials for more ample discussion than suits the nature of a historical outline. Such problems as that of finding the centre of oscillation,—the nature of the catenarian curve,—the determination of the line of swiftest descent,—the retardation produced to motion in a medium that resists according to the square of the velocity, or indeed according to any function of it,—the determination of the elastic curve, or that into which an elastic spring forms itself when a force is applied to bend it,—all these were problems of the greatest interest, and were now resolved for the first time; the science of mechanics being sufficient, by means of the composition of forces, to find out the fluxionary or differential equations which expressed the nature of the gradual changes which in all these cases were produced, and the calculus being now sufficiently powerful to infer the properties of the finite from those of the infinitesimal quantities.

The doctrine of Hydrostatics was cultivated in England by Cotes. The properties of the atmosphere, or of elastic fluids, were also experimentally investigated; and the barometer, after the ingenuity of Pascal had proved that the mercury stood lower the higher up into the atmosphere the instrument was carried, was at length brought to be a measure of the height of mountains. Mariotte appears to have been the first who proposed this use of it, and who discovered that, while the height from the ground increases in arithmetical, the density of the atmosphere, and the column of mercury in the barometer, decrease in geometrical progression. Halley, who seems also to have come of himself to the same conclusion, proved its truth by strict geometrical reasoning, and showed, that logarithms are easily applicable on this principle to the problem of finding the height of mountains. This was in the year 1685. Newton two years afterwards gave a demonstration of the same, extended to the case when gravity is not constant, but varies as any power of the distance from a given centre.

To the assiduous observations and the indefatigable activity of Halley, the natural history of the atmosphere, of the ocean, and of magnetism, are all under the greatest obligations. For the purpose of inquiring into these objects, this ardent and philosophical observer relinquished the quiet of academical retirement, and, having gone to St Helena, by a residence of a year in that island, not only made an addition to the catalogue of the stars, of 360 from the southern hemisphere, but returned with great acquisitions both of nautical and meteorological knowledge. His observations on evaporation were the foundation of two valuable papers on the origin of fountains; in which, for the first time, the sufficiency of the vapour taken up into the atmosphere, to maintain the perennial flow of springs and rivers, was established by undeniable evidence. The difficulty which men found in conceiving how a precarious and accidental supply like that of the rains, can sufficiently provide for a great and regular expenditure like that of the rivers, had given rise to those various opinions concerning the origin of fountains, which had hitherto divided the scientific world. A long residence on the summit of an insulated rock in the midst of a vast ocean, visited twice every year by the vertical sun, would have afforded to an observer, less quick-sighted than Halley, an opportunity of seeing the work of evaporation carried on with such rapidity and copiousness as to be a subject of exact measurement. From this extreme case, he could infer the medium quantity, at least by approximation; and he proved that, in the Mediterranean, the humidity daily raised up by evaporation is three times as great as that which is discharged by all the rivers that flow into it. The origin of fountains was no longer questioned, and of the multitude of opinions on that subject, which had hitherto perplexed philosophers, all but one entirely disappeared. *

Beside the voyage to St Helena, Halley made two others; the British government having been enlightened, and liberal enough to despise professional *etiquette*, where the interests of science were at stake, and to entrust to a Doctor of Laws the command of a ship of war, in which he traversed the Atlantic and Pacific Oceans in various directions, as far as the 53d degree of south latitude, and returned with a collection of facts and observations for the improvement of geography, meteorology, and navigation, far beyond that which any individual traveller or voyager had hitherto brought together.

The variation of the compass was long before this time known to exist, but its laws had never yet been ascertained. These Halley now determined from his own observations, combined with those of former navigators, in so far as to trace, on a nautical chart, the lines of the same variation over a great part both of the Atlantic and Pacific Oceans, affording to the navigator the ready means of correcting the errors which the deviation of the needle, from the true meridian, was calculated to produce. In his different traverses he had four times intersected the line of no variation, which

* Philosophical Transactions, 1687, Vol. XVI. p. 366.

seemed to divide the earth into two parts, the variations on the east side being towards the west, and on the west side towards the east. These lines being found to change their position in the course of time, the place assigned to the magnetical poles could not be permanent. Any theory, therefore, which could afford an explanation of their changes must necessarily be complex and difficult to be established. The attempt of Halley to give such an explanation, though extremely ingenious, was liable to great objections; and while it has shared the fate of most of the theories which have been laid down before the phenomena had been sufficiently explored, the general facts which he established have led to most of the improvements and discoveries which have since been made respecting the polarity of the needle.

Besides the conclusion just mentioned, Dr Halley derived, from his observations, a very complete history of the winds which blow in the tropical regions, viz. the trade-wind, and the monsoons, together with many interesting facts concerning the phenomena of the tides. The chart which contained an epitome of all these facts was published in 1701.

The above are only a part of the obligations which the sciences are under to the observations and reasonings of this ingenious and indefatigable inquirer. Halley was indeed one of the ablest and most accomplished men of his age. A scholar well versed in the learned languages, and a geometer profoundly skilled in the ancient analysis, he restored to their original elegance some of the precious fragments of that analysis, which time happily had not entirely defaced. He was well acquainted also with the algebraical and fluxionary calculus, and was both in theory and practice a profound and laborious astronomer. Finally, he was the friend of Newton, and often stimulated, with good effect, the tardy purposes of that great philosopher. Few men, therefore, of any period, have more claims than Halley on the gratitude of succeeding ages.

The invention of the thermometer has been already noticed, and the improvements made on that instrument about this period, laid the foundation of many future discoveries. The discovery of two fixed temperatures, each marked by the same expansion of the mercury in the thermometer, and the same condition of the fluid in which it is immersed, was made about this time. The differences of temperature were thus subjected to exact measurement; the phenomena of heat became, of course, known with more certainty and precision; and that substance or virtue, to which nothing is impenetrable, and which finds its way through the rarest and the densest bodies, apparently with the same facility,—which determines so many of our sensations, and of which the distribution so materially influences all the phenomena of animal and vegetable life, came now to be known, not, indeed, in its essence, but as to all the characters in which we are practically or experimentally concerned. The treatise on Fire, in Boerhaave's Chemistry, is a great advance beyond any thing on that subject hitherto known, and touches, notwithstanding many errors and imperfections, on most of the great truths, which time, experience, and ingenuity, have since brought into view.

It was in this period also that electricity may be said first to have taken a scientific form. The power of amber to attract small bodies, after it has been rubbed, is said to have been known to Thales, and is certainly made mention of by Theophrastus. The observations of Gilbert, a physician of Colchester, in the end of the sixteenth century, though at the distance of two thousand years, made the first addition to the transient and superficial remarks of the Greek naturalist, and afford a pretty full enumeration of the bodies which can be rendered electrical by friction. The Academia del Cimento, Boyle, and Otto Guericke, followed in the same course; and the latter is the first who mentions the crackling noise and faint light which electricity sometimes produced. These, however, were hardly perceived, and it was by Dr Wall, as described in the Philosophical Transactions, that they were first distinctly observed.* By a singularly fortunate anticipation, he remarks of the light and crackling, that they seemed, in some degree, to represent thunder and lightning.

After the experiments of Hauksbee in 1709, by which the knowledge of this mysterious substance was considerably advanced, Wheeler and Gray, who had discovered that one body could communicate electricity to another without rubbing, being willing to try to what distance the electrical virtue might be thus conveyed, employed, for the purpose of forming the communication, hempen rope, which they extended to a considerable length, supporting it from the sides, by threads which, in order to prevent the dissipation of the electricity, they thought it proper to make as slender as possible. They employed silk threads with that view, and found the experiment to succeed. Thinking that it would succeed still better, if the supports were made still more slender, they tried very fine metallic wire, and were surprised to find, that the hempen rope, thus supported, conveyed no electricity at all. It was, therefore, as being *silk*, and not as being *small*, that the threads had served to retain the electricity. This accident

* Wall's paper is in the *Transactions* for 1708, Vol. XXVI. No. 314, p. 69.---Hauksbee on *Electrical Light*, in the same volume. See *Abridgment*, Vol. V. p. 408, 411.

led to the great distinction of substances conducting and not conducting electricity. An extensive field of inquiry was thus opened, a fortunate accident having supplied an *instantia crucis*, and enabled these experimenters to distinguish between what was essential and what was casual in the operation they had performed. The history of electricity, especially in its early stages, abounds with facts of this kind; and no man, who would study the nature of inductive science, and the rules for the interpretation of nature, can employ himself better than in tracing the progress of these discoveries. He will find abundant reason to admire the ingenuity as well as the industry of the inquirers, but he will often find accident come in very opportunely to the assistance of both. The experiments of Wheeler and Gray are described in the Transactions for 1729.

SECTION III.

OPTICS.

THE invention of the telescope and the microscope, the discoveries made concerning the properties of light and the laws of vision, added to the facility of applying mathematical reasoning as an instrument of investigation, had long given a peculiar interest to optical researches. The experiments and inquiries of Newton on that subject began in 1666, and soon made a vast addition both to the extent and importance of the science. He was at that time little more than twenty-three years old; he had already made some of the greatest and most original discoveries in the pure mathematics; and the same young man, whom we have been admiring as the most profound and inventive of geometers, is to appear, almost at the same moment, as the most patient, faithful, and sagacious interpreter of nature. These characters, though certainly not opposed to one another, are not often combined; but to be combined in so high a degree, and in such early life, was hitherto without example.

In hopes of improving the telescope, by giving to the glasses a figure different from the spherical, he had begun to make experiments, and had procured a glass prism, in order, as he tells us, to try with it the *celebrated phenomena of colours.* * These trials led to the discovery of the different refrangibility of the rays of light, and are now too well known to stand in need of a particular description.

Having admitted a beam of light into a dark chamber, through a hole in the window-shutter, and made it fall on a glass prism, so placed as to cast it on the opposite wall, he was delighted to observe the brilliant colouring of the sun's image, and not less surprised to observe its figure, which, instead of being circular, as he expected, was oblong in the direction perpendicular to the edges of the prism, so as to have the shape of a parallelogram, rounded at the two ends, and nearly five times as long as it was broad.

When he reflected on these appearances, he saw nothing that could explain the elongation of the

---

* *Phil. Trans.* Vol. VI. (1672,) p. 3075. Also Hutton's *Abridgment,* Vol. I. p. 678. The account of the experiments is in a letter to Oldenburg, dated February 1672; it is the first of Newton's works that was published. It is plain from what is said above, that the phenomena of the prismatic spectrum were not unknown at that time, however little they were understood, and however imperfectly observed.

image but the supposition that some of the rays of light, in passing through the prism, were more refracted than others, so that rays which were parallel when they fell on the prism, diverged from one another after refraction, the rays that differed in refrangibility differing also in colour. The *spectrum*, or solar image, would thus consist of a series of circular images partly covering one another, and partly projecting one beyond another, from the red or least refrangible rays, in succession, to the orange, yellow, green, blue, indigo, and violet, the most refrangible of all.

It was not, however, till he tried every other hypothesis which suggested itself to his mind by the test of experiment, and proved its fallacy, that he adopted this as a true interpretation of the phenomena. Even after these rejections, his explanation had still to abide the sentence of an *experimentum crucis*.

Having admitted the light and applied a prism as before, he received the coloured spectrum on a board at the distance of about twelve feet from the first, and also pierced with a small hole. The coloured light which passed through this second hole was made to fall on a prism, and afterwards received on the opposite wall. It was then found that the rays which had been most refracted, or most bent from their course by the first prism, were most refracted also by the second, though no new colours were produced. "So," says he, "the true cause of the length of the image was detected to be no other than that light consists of rays differently refrangible, which, without any respect to a difference in their incidence, were, according to their degrees of refrangibility, transmitted towards divers parts of the wall." *

It was also observed, that when the rays which fell on the second prism were all of the same colour, the image formed by refraction was truly circular, and of the same colour with the incident light. This is one of the most conclusive and satisfactory of all the experiments.

When the sun's light is thus admitted first through one aperture, and then through another at some distance from the first, and is afterwards made to fall on a prism, as the rays come only from a part of the sun's disk, the spectrum has nearly the same length as before, but the breadth is greatly diminished; in consequence of which, the light at each point is purer, it is free from penumbra, and the confines of the different colours can be more accurately traced. It was in this way that Newton measured the extent of each colour, and taking the mean of a great number of measures, he assigned the following proportions, dividing the whole length

* Phil. Trans. Vol. VI. (1672,) No. 80. p. 3075.

of the spectrum, exclusive of its rounded terminations, into 360 equal parts; of these the

Red occupied	-	45
Orange	- -	27
Yellow	- -	48
Green	- -	60
Blue	- -	60
Indigo	- -	40
Violet	- -	80.

Between the divisions of the spectrum, thus made by the different colours, and the divisions of the monochord by the notes of music, Newton conceived that there was an analogy, and indeed an identity of ratios; but experience has since shown that this analogy was accidental, as the spaces occupied by the different colours do not divide the spectrum in the same ratio, when prisms of different kinds of glass are employed.

Such were the experiments by which Newton first " untwisted all the shining robe of day," and made known the texture of the magic garment which nature has so kindly spread over the surface of the visible world. From them it followed, that colours are not qualities which light derives from refraction or reflection, but are original and *connate* properties connected with the different degrees of refrangibility that belong to the different rays. The same colour is always joined to the same degree of refrangibility, and conversely, the same degree of refrangibility to the same colour.

Though the seven already enumerated are primary and simple colours, any of them may also be produced by a mixture of others. A mixture of yellow and blue, for instance, makes green; of red and yellow orange; and, in general, if two colours, which are not very far asunder in the natural series, be mixed together, they compound the colour that is in the middle between them.

But the most surprising composition of all, Newton observes, is that of whiteness; which is not produced by one sort of rays, but by the mixture of all the colours in a certain proportion, namely, in that proportion which they have in the solar spectrum. This fact may be said to be made out both by analysis and composition. The white light of the sun can be separated, as we have just seen, into the seven simple colours; and if these colours be united again they form white. Should any of them have been wanting, or not in its due proportion, the white produced is defective.

It appeared, too, that natural bodies, of whatever colour, if viewed by simple and homogeneous light, are seen of the colour of that light, and of no other. Newton tried this very satisfactory experiment on

4

bodies of all colours, and found it to hold uniformly; the light was never changed by the colour of the body that reflected it.

Newton, thus furnished with so many new and accurate notions concerning the nature and production of colour, proceeded to apply them to the explanation of phenomena. The subject which naturally offered itself the first to this analysis was the rainbow, which, by the grandeur and simplicity of its figure, added to the brilliancy of its colours, in every age has equally attracted the attention of the peasant and of the philosopher. That two refractions and one reflection were at least a part of the machinery which nature employed in the construction of this splendid arch, had been known from the time of Antonio de Dominis; and the manner in which the arched figure is produced had been shown by Descartes; so that it only remained to explain the nature of the colour and its distribution. As the colours were the same with those exhibited by the prism, and succeeded in the same order, it could hardly be doubted that the cause was the same. Newton showed the truth of his principles by calculating the extent of the arch, the breadth of the coloured bow, the position of the secondary bow, its distance from the primary, and by explaining the inversion of the colours. * There is not, perhaps,

---

Optics, Book I. Prop. 9.

in science any happier application of theory, or any in which the mind rests with fuller confidence.

Other meteoric appearances seemed to be capable of similar explanations, but the phenomena being no where so regular or so readily subjected to measurement as those of the rainbow, the theory cannot be brought to so severe a test, nor the evidence rendered so satisfactory.

But a more difficult task remained,—to explain the permanent colour of natural bodies. Here, however, as it cannot be doubted that all colour comes from the rays of light, so we must conclude that one body is red and another violet, because the one is disposed to reflect the red or least refrangible rays, and the other to reflect the violet or the most refrangible. Every body manifests its disposition to reflect the light of its own peculiar colour, by this, that if you cast on it pure light, first of its own colour, and then of any other, it will reflect the first much more copiously than the second. If cinnabar, for example, and ultra-marine blue be both exposed to the same red homogeneous light, they will both appear red ; but the cinnabar strongly luminous and resplendent, and the ultra-marine of a faint obscure red. If the homogeneal light thrown on them be blue, the converse of the above will take place.

Transparent bodies, particularly fluids, often transmit light of one colour and reflect light of another. Halley told Newton, that, being deep under the surface of the sea in a diving-bell, in a clear sunshine day, the upper side of his hand, on which the sun shone darkly through the water, and through a small glass window in the diving-bell, appeared of a red colour, like a damask rose, while the water below, and the under part of his hand, looked green. *

But, in explaining the permanent colour of bodies, this difficulty always presents itself,—Suppose that a body reflects red or green light, what is it that decomposes the light, and separates the red or the green from the rest? Refraction is the only means of decomposing light, and separating the rays of one degree of refrangibility and of one colour, from those of another. This appears to have been what led Newton to study the colours produced by light passing through thin plates of any transparent substance. The appearances are very remarkable, and had already attracted the attention both of Boyle and of Hooke, but the facts observed by them remained insulated in their hands, and unconnected with other optical phenomena.

It probably had been often remarked, that when two transparent bodies, such as glass, of which the surfaces were convex in a certain degree, were pressed together, a black spot was formed at the contact

---

* Optics, p. 115. Horsley's edit.

of the two, which was surrounded with coloured rings, more or less regular, according to the form of the surfaces. In order to analyze a phenomenon that seemed in itself not a little curious, Newton proposed to make the experiment with surfaces of a regular curvature, such as was capable of being measured. He took two object-glasses, one a plano-convex for a fourteen feet telescope, the other a double convex for one of about fifty feet, and upon this last he laid the other with its plane side downwards, pressing them gently together. At their contact in the centre was a pellucid spot, through which the light passed without suffering any reflection. Round this spot was a coloured circle or ring, exhibiting blue, white, yellow, and red. This was succeeded by a pellucid or dark ring, then a coloured ring of violet, blue, green, yellow, and red, all copious and vivid except the green. The third coloured ring consisted of purple, blue, green, yellow, and red. The fourth consisted of green and red; those that succeeded became gradually more dilute, and ended in whiteness. It was possible to count as far as seven.

The colours of these rings were so marked by peculiarities in shade and vivacity, that Newton considered them as belonging to different orders; so that an eye accustomed to examine them, on any particular colour of a natural object being pointed out, would be able to determine to what order in this series it belonged.

Thus we have a system of rings or zones surrounding a dark central spot, and themselves alternately dark and coloured, that is, alternately transmitting the light and reflecting it. It is evident that the thickness of the plates of air interposed between the glasses, at each of those rings, must be a very material element in the arrangement of this system. Newton, therefore, undertook to compute their thickness. Having carefully measured the diameters of the first six coloured rings, at the most lucid part of each, he found their squares to be as the progression of odd numbers 1, 3, 5, 7, &c. The squares of the distances from the centre of the dark spot to each of these circumferences, were, therefore, in the same ratio, and consequently the thickness of the plates of air, or the intervals between the glasses, were as the numbers 1, 3, 5. 7, &c.

When the diameters of the dark or pellucid rings which separated the coloured rings were measured, their squares were found to be as the even numbers 0, 2, 4, 6, and, therefore, the thickness of the plates through which the light was wholly transmitted were as the same numbers. A great many repeated measurements assured the accuracy of these determinations.

As the curvature of the convex glass on which

the flat surface of the plano-convex rested was known, and as the diameters of the rings were measured in inches, it was easy to compute the thickness of the plates of air, which corresponded to the different rings.

An inch being divided into 178000 parts, the distance of the lenses for the first series, or for the luminous rings, was $\frac{1}{178000}$, $\frac{3}{178000}$, $\frac{5}{178000}$, &c.

For the second series $\frac{2}{178000}$, $\frac{4}{178000}$, &c.

When the rings were examined by looking through the lenses in the opposite direction, the central spot appeared white, and, in other rings, red was opposite to blue, yellow to violet, and green to a compound of red and violet; the colours formed by the transmitted and the reflected light being, what is now called complementary, or nearly so, of one another; that is, such as when mixed produce white.

When the fluid between the glasses was different from air, as when it was water, the succession of rings was the same; the only difference was, that the rings themselves were narrower.

When experiments on thin plates were made in such a way that the plate was of a denser body than the surrounding medium, as in the case of soap-bubbles, the same phenomena were observed to take place. These phenomena Newton also examined with his accustomed accuracy, and even bestowed particular care on having the soap-bubbles as perfect and durable as their frail structure would admit. In the eye of philosophy no toy is despicable, and no occupation frivolous, that can assist in the discovery of truth.

To the different degrees of tenuity, then, in transparent substances, there seemed to be attached the powers of separating particular colours from the mass of light, and of rendering them visible sometimes by reflection, and, in other cases, by transmission. As there is reason to think, then, that the minute parts, the mere particles of all bodies, even the most opaque, are transparent, they may very well be conceived to act on light after the manner of the thin plates, and to produce each, according to its thickness and density, its appropriate colour, which, therefore, becomes the colour of the surface. Thus the colours in which the bodies round us appear everywhere arrayed, are reducible to the action of the parts which constitute their surfaces on the refined and active fluid which pervades, adorns, and enlightens the world.

But the same experiments led to some new and unexpected conclusions, that seemed to reach the very essence of the fluid of which we now speak. It was impossible to observe, without wonder, the rings alternately luminous and dark that were formed between the two plates of glass in the preceding experiments, and determined to be what they were by the different thickness of the air between the plates, and having to that thickness the relations formerly expressed. A plate of which the thickness was equal to a certain quantity multiplied by an odd number, gave always a circle of the one kind; but if the thickness of the plate was equal to the same quantity multiplied by an even number, the circle was of another kind, the light, in the first case, being reflected, in the second transmitted. Light penetrating a thin transparent plate, of which the thickness was m, $3m$, $5m$, &c. was decomposed and reflected; the same light penetrating the same plate, but of the thickness 0, $2m$, $4m$, was transmitted, though, in a certain degree, also decomposed. The same light, therefore, was transmitted or reflected, according as the second surface of the plate of air through which it passed was distant from the first by the intervals 0, $2m$, $4m$, or m, $3m$, $5m$; so that it becomes necessary to suppose the same ray to be successively disposed to be transmitted and to be reflected at points of space separated from one another by the same interval m. This constitutes what Newton called *Fits of easy transmission and easy reflection*, and forms one of the most singular parts of his optical discoveries. It is so unlike any thing which analogy teaches us to expect, that it has often been viewed with a degree of incredulity, and regarded as at best but a conjecture introduced to account for certain optical phenomena. This, however, is by no means a just conclusion, for it is, in reality, a necessary inference from appearances accurately observed, and is no less entitled to be considered as a fact than those appearances themselves. The difficulty of assigning a cause for such extraordinary alternations cannot be denied, but does not entitle us to doubt the truth of a conclusion fairly deduced from experiment. The principle has been confirmed by phenomena that were unknown to Newton himself, and possesses this great and unequivocal character of philosophic truth, that it has served to explain appearances which were not observed till long after the time when it first became known.

We cannot follow the researches of Newton into what regards the colours of thick plates, and of bodies in general. We must not, however, pass over his explanation of refraction, which is among the happiest to be met with in any part of science, and has the merit of connecting the principles of optics with those of dynamics.

The theory from which the explanation we speak of is deduced is, that light is an emanation of particles, moving in straight lines with incredible velocity, and attracted by the particles of transparent bodies. When, therefore, light falls obliquely on the surface of such a body, its motion may be resolved into two, one parallel to that surface, and the other perpendicular to it. Of these, the first is not affected by the attraction of the body, which is perpendicular to its own surface; and, therefore, it remains the same in the refracted that it was in the incident ray. But the velocity perpendicular to the surface is increased by the attraction of the body, and, according to the principles of dynamics, (the 39th, Book I. Princip.) whatever be the quantity of this velocity, its square, on entering the same transparent body, will always be augmented by the same quantity. But it is easy to demonstrate that, if there be two right-angled triangles, with a side in the one equal to a side in the other, the hypothenuse of the first being given, and the squares of their remaining sides differing by a given space, the sines of the angles opposite to the equal sides must have a given ratio to one another.* This amounts to the same with saying, that, in the case before us, the sine of the angle of incidence is to the sine of the angle of refraction in a given ratio. The explanation of the law of refraction thus given is so highly satisfactory, that it affords a strong argument in favour of the system which considers light as an emanation of particles from luminous bodies, rather than the vibrations of an elastic fluid. It is true that Huygens deduced

---

* Optics, Book II. Part iii. Prop. 10.

from this last hypothesis an explanation of the law of refraction, on which considerable praise was bestowed in the former part of this Dissertation. It is undoubtedly very ingenious, but does not rest on the same solid and undoubted principles of dynamics with the preceding, nor does it leave the mind so completely satisfied. Newton, in his Principia, has deduced another demonstration of the same optical proposition from the theory of central forces. *

The different refrangibility of the rays of light forms no exception to the reasoning above. The rays of each particular colour have their own particular ratio subsisting between the sines of incidence and refraction, or in each, the square that is added to the square of the perpendicular velocity has its own value, which continues the same while the transparent medium is the same.

Light, in consequence of these views, became, in the hands of Newton, the means of making important discoveries concerning the internal and chemical constitution of bodies. The square that is added to that of the perpendicular velocity of light in consequence of the attractive force of the transparent substance, is properly the measure of the quantity of that attraction, and is

* Prin. Math. Lib. I. Prop. 94. Also Optics, Book I. Prop. 6.

the same with the difference of the squares of the velocities of the incident and the refracted light. This is readily deduced, therefore, from the ratio of the angle of incidence to that of refraction ; and when this is done for different substances, it is found, that the above measure of the refracting power of transparent bodies is nearly proportional to their density, with the exception of those which contain much inflammable matter in their composition, or sulphur, as it was then called, which is always accompanied with an increase of refracting power.*

Thus, the refracting power, ascertained as above, when divided by the density, gives quotients not very different from one another, till we come to the inflammable bodies, where a great increase immediately takes place. In air, for instance, the quotient is 5208, in rock-crystal 5450, and the same nearly in common glass. But in spirit of wine, oil, amber, the same quotients are 10121, 12607, 13654. Newton found in the diamond, that this quotient is still greater than any of the preceding, being 14556.† Hence he conjectured, what has since been so fully verified by experiment, that the diamond, at least in part, is an inflammable body. Observing, also, that the refracting power of water is great for its density, the quo-

* Newton's Optics, Ibid. † Ibid.

tient, expounding it as above, being 7845, he concluded, that an inflammable substance enters into the composition of that fluid,—a conclusion which has been confirmed by one of the most certain but most unexpected results of chemical analysis. The views thus suggested by Newton have been successfully pursued by future inquirers, and the action of bodies on light is now regarded as one of the means of examining into their internal constitution.

I should have before remarked, that the alternate disposition to be easily reflected and easily transmitted, serves to explain the fact, that all transparent substances reflect a portion of the incident light. The reflection of light from the surfaces of opaque bodies, and from the anterior surfaces of transparent bodies, appears to be produced by a repulsive force exerted by those surfaces at a determinate but very small distance, in consequence of which there is stretched out over them an elastic web through which the particles of light, notwithstanding their incredible velocity, are not always able to penetrate.* In the case of a transparent body, the light which, when it arrives at this outwork, as it may be called, is in a *fit* of easy reflec-

* A velocity that enables light to pass from the sun to the earth in 8′ 13″, as is deduced from the eclipses of Jupiter's satellites.

tion, obeys of course the repulsive force, and is reflected back again. The particles, on the other hand, which are in the state which disposes them to be transmitted, overcome the repulsive force, and, entering into the interior of the transparent body, are subjected to the action of its attractive force, and obey the law of refraction already explained. If these rays, however, reach the second surface of the transparent body, (that body being supposed denser than the medium surrounding it,) in a direction having a certain obliquity to that surface, the attraction will not suffer the rays to emerge into the rarer medium, but will force them to return back into the transparent body. Thus the reflection of light at the second surface of a transparent body is produced, not by the repulsion of the medium in which it was about to enter, but by the attraction of that which it was preparing to leave.

The first account of the experiments from which all these conclusions were deduced, was given in the Philosophical Transactions for 1672, and the admiration excited by their brilliancy and their novelty may easily be imagined. Among the men of science, the most enlightened were the most enthusiastic in their praise. Huygens, writing to one of his friends, says of them, and of the truths they were the means of making known, "*Quoram respectu omnia huc usque edita jejunia sunt et pror-*

sus puerilia." Such were the sentiments of the person who, of all men living, was the best able to judge, and had the best right to be fastidious in what related to optical experiments and discoveries. But all were not equally candid with the Dutch philosopher; and though the discovery now communicated had every thing to recommend it which can arise from what is great, new and singular; though it was not a theory or a system of opinions, but the generalization of facts made known by experiments; and though it was brought forward in the most simple and unpretending form, a host of enemies appeared, each eager to obtain the unfortunate pre-eminence of being the first to attack conclusions which the unanimous voice of posterity was to confirm. In this contention, the envy and activity of Hooke did not fail to give him the advantage, and he communicated his objections to Newton's conclusions concerning the refrangibility of light in less than a month after they had been read in the Royal Society. He admitted the accuracy of the experiments themselves, but denied that the cause of the colour is any quality residing permanently in the rays of light, any more than that the sounds emitted from the pipes of an organ exist originally in the air. An imaginary analogy between sound and light seems to have been the basis of all his optical theories. He conceived that colour is nothing but the disturbance of light by pulses propagated through it; that blackness proceeds from the scarcity, whiteness from the plenty, of undisturbed light; and that the prism acts by exciting different pulses in this fluid, which pulses give rise to the sensations of colour. This obscure and unintelligible theory (if we may honour what is unintelligible with the name of a theory) he accompanied with a multitude of captious objections to the reasonings of Newton, whom he was not ashamed to charge with borrowing from him without acknowledgment. To all this Newton replied, with the solidity, calmness, and modesty, which became the understanding and the temper of a true philosopher.

The new theory of colours was quickly assailed by several other writers, who seem all to have had a better apology than Hooke for the errors into which they fell. Among them one of the first was Father Pardies, who wrote against the experiments, and what he was pleased to call the hypothesis, of Newton. A satisfactory and calm reply convinced him of his mistake, which he had the candour very readily to acknowledge. A countryman of his, Mariotte, was more difficult to be reconciled, and, though very conversant with experiment, appears never to have succeeded in repeating the experiments of Newton. Desaguliers, at the request of the latter, repeated the experiments doubted of before the Royal Society, where Monmort, a countryman and a friend of Mariotte, was present. *

MM. Linus and Lucas, both of Liege, objected to Newton's experiments as inaccurate; the first, because, on attempting to repeat them, he had not obtained the same results; and the second, because he had not been able to perceive that a red object and a blue required the focal distance to be different when they were viewed through a telescope. Newton replied with great patience and good temper to both.

The series was closed, in 1727, by the work of an Italian author, Rizetti, who, in like manner, called in question the accuracy of experiments which he himself had not been able to repeat. Newton was now no more, but Desaguliers, in consequence of Rizetti's doubts, instituted a series of experiments which seemed to set the matter entirely at rest. These experiments are described in the Philosophical Transactions for 1728.

An inference which Newton had immediately drawn from the discoveries above described was, that the great source of imperfection in the refracting telescope was the different refrangibility of the rays of light, and that there were stronger reasons than either Mersenne or Gregory had suspected, for looking to reflection for the improvement of

* Montucla, Tom. II.

optical instruments. It was evident, from the different refrangibility of light, that the rays coming from the same point of an object, when decomposed by the refraction of a lens, must converge to different foci; the red rays, for example, to a point more distant from the lens, and the violet to one nearer by about a fifty-fourth part of the focal distance. Hence it was not merely from the aberration of the rays caused by the spherical figure of the lens that the imperfection of the images formed by refraction arose, but from the very nature of refraction itself. It was evident, at the same time, that in a combination of lenses with opposite figures, one convex, for instance, and another concave, there was a tendency of the two contrary dispersions to correct one another. But it appeared to Newton, on examining different refracting substances, that the dispersion of the coloured rays never could be corrected except when the refraction itself was entirely destroyed, for he thought he had discovered that the quantity of the refraction and of the dispersion in different substances bore always the same proportion to one another. This is one of the few instances in which his conclusions have not been confirmed by subsequent experiment; and it will, accordingly, fall under discussion in another part of this discourse.

Having taken the resolution of constructing a reflecting telescope, he set about doing so with his own hands. There was, indeed, at that time, no other means by which such a work could be accomplished; the art of the ordinary glass-grinder not being sufficient to give to metallic specula the polish which was required. It was on this account that Gregory had entirely failed in realizing his very ingenious optical invention.

Newton, however, himself possessed excellent hands for mechanical operations, and could use them to better purpose than is common with men so much immersed in deep and abstract speculation. It appears, indeed, that mechanical invention was one of the powers of his mind which began to unfold itself at a very early period. In some letters subjoined to a Memoir drawn up after his death by his nephew Conduit, it is said, that, when a boy, Newton used to amuse himself with constructing machines, mills, &c. on a small scale, in which he displayed great ingenuity; and it is probable that he then acquired that use of his hands which is so difficult to be learned at a later period. To this, probably, we owe the neatness and ingenuity with which the optical experiments above referred to were contrived and executed,—experiments of so difficult a nature, that any error in the manipulation would easily defeat the effect, and appears actually to have done so with many of those who objected to his experiments. *

* The Memoir of Conduit was sent to Fontenelle when

He succeeded perfectly in the construction of his telescope, and his first communication with Oldenburg, and the first reference to his optical

---

he was preparing the *Eloge* on Newton, but he seems to have paid little attention to it, and has passed over the early part of his life with the remark, that one may apply to him what Lucan says of the Nile, that it has not been "permitted to mortals to see that river in a feeble state." If the letters above referred to had formed a part of this communication, I think the Secretary of the Academy would have sacrificed a fine comparison to an instructive fact. In other respects Conduit's Memoir did not convey much information that could be of use. His instructions to Fontenelle are curious enough; he bids him be sure to state, that Leibnitz had borrowed the Differential Calculus from the Method of Fluxions. He conjured him in another place not to omit to mention, that Queen Caroline used to delight much in the conversation of Newton, and nothing could do more honour to Newton than the commendation of a Queen, the Minerva of her age. Fontenelle was too much a philosopher, and a man of the world, (and had himself approached too near to the persons of princes,) to be of Mr Conduit's opinion, or to think that the approbation of the most illustrious princess could add dignity to the man who had made the three greatest discoveries yet known, and in whose hands the sciences of Geometry, Optics, and Astronomy, had all taken new forms. If he had been called to write the *Eloge* of the Queen of England, he would, no doubt, have remarked her relish for the conversation of Newton.

On the whole, the *Eloge* on Newton has great merit, and, to be the work of one who was at bottom a Cartesian, is a singular example of candour and impartiality.

experiments, is connected with the construction of this instrument, and mentioned in a letter dated the 11th January 1672. He had then been proposed as a member of the Royal Society by the Bishop of Sarum, and he says, "If the honour of being a member of the Society shall be conferred on me, I shall endeavour to testify my gratitude by communicating what my poor and solitary endeavours can effect toward the promoting its philosophical designs." * Such was the modesty of the man who was to effect a greater revolution in the state of our knowledge of nature than any individual had yet done, and greater, perhaps, than any individual is ever destined to bring about. Success, however, never altered the temper in which he began his researches.

Newton, after considering the reflection and refraction of light, proceeded, in the third and last Book of his Optics, to treat of its *inflection*, a subject which, as has been remarked in the former part of this discourse, was first treated of by Grimaldi. Newton having admitted a ray of light through a hole in a window-shutter into a dark chamber, made it pass by the edge of a knife, or, in some experiments, between the edges of two knives, fixed parallel, and very near to one another; and, by receiving the light on a sheet of paper at

* Birch's History of the Royal Society, Vol. III. p. 3.

different distances behind the knives, he observed the coloured fringes which had been described by the Italian optician, and, on examination, found, that the rays had been acted on in passing the knife edges both by repulsive and attractive forces, and had begun to be so acted on in a sensible degree when they were yet distant by $\frac{1}{800}$ of an inch from the edges of the knives. His experiments, however, on this subject were interrupted, as he informs us, and do not appear to have been afterwards resumed. They enabled him, however, to draw this conclusion, that the path of the ray in passing by the knife edge was bent in opposite directions, so as to form a serpentine line, convex and concave toward the knife, according to the repulsive or attractive forces which acted at different distances; that it was also reasonable to conclude, that the phenomena of the refraction, reflection, and inflection of light, were all produced by the same force variously modified, and that they did not arise from the actual contact or collision of the particles of light with the particles of bodies.

The Third Book of the Optics concludes with those celebrated Queries which carry the mind so far beyond the bounds of ordinary speculation, though still with the support and under the direction either of direct experiment or close analogy. They are a collection of propositions relative chiefly to the nature of the mutual action of light and of bodies on one another, such as appeared to the author highly probable, yet wanting such complete evidence as might entitle them to be admitted as principles established. Such enlarged and comprehensive views, so many new and bold conceptions, were never before combined with the sobriety and caution of philosophical induction. The anticipation of future discoveries, the assemblage of so many facts from the most distant regions of human research, all brought to bear on the same points, and to elucidate the same questions, are never to be sufficiently admired. At the moment when they appeared, they must have produced a wonderful sensation in the philosophic world, unless, indeed, they advanced too far before the age, and contained too much which the comment of time was yet required to elucidate.

It is in the Queries that we meet with the ideas of this philosopher concerning the *Elastic Ether*, which be conceived to be the means of conveying the action of bodies from one part of the universe to another, and to which the phenomena of light, of heat, of gravitation, are to be ascribed. Here we have his conclusions concerning that polarity or peculiar virtue residing in the opposite sides of the rays of light, which he deduced from the enigmatical phenomena of doubly refracting crystals. Here, also, the first step is made toward the doctrine of elective attractions or of chemical affinity, and to the notion, that the phenomena of chemistry, as well as of cohesion, depend on the alternate attractions and repulsions existing between the particles of bodies at different distances. The comparison of the gradual transition from repulsion to attraction at those distances, with the positive and negative quantities in algebra, was first suggested here, and is the same idea which the ingenuity of Boscovich afterwards expanded into such a beautiful and complete system. Others who have attempted such flights had ended in mere fiction and romance; it is only for such men as Bacon or Newton to soar beyond the region of poetical fiction, still keeping sight of probability, and alighting again safe on the *terra firma* of philosophic truth. *

* The optical works of Newton are not often to be found all brought together into one body. The first part of them consists of the papers in the Philosophical Transactions, which gave the earliest account of his discoveries, and which have been already referred to. They are in the form of Letters to Oldenburg, the Secretary of the Society, as are also the answers, to Hooke, and the others who objected to these discoveries; the whole forming a most interesting and valuable series which Dr Horsley has published in the fourth volume of his edition of Newton's works, under the title of Letters relating to the Theory of Light and Colours. The next work, in point of time, consists of the Lectiones Opticæ, or the optical lectures which the author delivered at Cambridge. The Optics, in three books, is the last and

SECTION IV.

ASTRONOMY.

THE time was now come when the world was to be enlightened by a new science, arising out of the comparison of the phenomena of motion as observed in the heavens, with the laws of motion as known on the earth. Physical astronomy was the result of this comparison, a science embracing greater objects, and destined for a higher flight than any other branch of natural knowledge. It is unnecessary to observe, that it was by Newton that the comparison just referred to was instituted, and the riches of the new science unfolded to mankind.

This young philosopher, already signalized by great discoveries, had scarcely reached the age of twenty-four, when a great public calamity forced him into the situation where the first step in the new science is said to have been suggested; and that, by some of those common appearances in

most complete, containing all the reasoning concerning optical phenomena above referred to. The first edition was in 1704, the second, with additions, in 1717. Newtoni Opera, Tom. IV. Horsley's edition.

which an ordinary man sees nothing to draw his attention, nor even the man of genius, except at those moments of inspiration when the mind sees farthest into the intellectual world. In 1666, the plague forced him to retire from Cambridge into the country ; and, as he sat one day alone, in a garden, musing on the nature of the mysterious force by which the phenomena at the earth's surface are so much regulated, he observed the apples falling spontaneously from the trees, and the thought occurred to him, since gravity is a tendency not confined to bodies on the very surface of the earth, but since it reaches to the tops of trees, to the tops of the highest buildings, nay, to the summits of the most lofty mountains, without its intensity or direction suffering any sensible change, Why may it not reach to a much greater distance, and even to the moon itself? And, if so, may not the moon be retained in her orbit by gravity, and forced to describe a curve like a projectile at the surface of the earth? *

Here another consideration very naturally occurred. Though gravity be not sensibly weakened at the small distances from the surface to which our experiments extend, it may be weakened at greater distances, and at the moon may be greatly diminished. To estimate the quantity of

* Pemberton's View of Newton's Philosophy, Pref.

this diminution Newton appears to have reasoned thus: If the moon be retained in her orbit by her gravitation to the earth, it is probable that the planets are, in like manner, carried round the sun by a power of the same kind with gravity, directed to the centre of that luminary. He proceeded, therefore, to inquire, by what law the tendency, or gravitation of the planets to the sun must diminish, in order that, describing, as they do, orbits nearly circular round the sun, their times of revolution and their distances may have the relation to one another which they are known to have from observation, or from the third law of Kepler.

This was an investigation which, to most even of the philosophers and mathematicians of that age, would have proved an insurmountable obstacle to their farther progress; but Newton was too familiar with the geometry of evanescent or infinitely small quantities, not to discover very soon, that the law now referred to would require the force of gravity to diminish exactly as the square of the distance increased. The moon, therefore, being distant from the earth about sixty semidiameters of the earth, the force of gravity at that distance must be reduced to the 3600th part of what it is at the earth's surface. Was the deflection of the moon then from the tangent of her orbit, in a second of time, just the 3600th part of the distance which a heavy body falls in a second at the surface of the earth? This was a question that could be precisely answered, supposing the moon's distance known not merely in semidiameters of the earth but in feet, and her angular velocity, or the time of her revolution in her orbit, to be also known.

In this calculation, however, being at a distance from books, he took the common estimation of the earth's circumference that was in use before the measurement of Norwood, or of the French Academicians, according to which, a degree is held equal to 60 English miles. This being in reality a very erroneous supposition, the result of the calculation did not represent the force as adequate to the supposed effect; whence Newton concluded that some other cause than gravity must act on the moon, and on that account he laid aside, for the time, all farther speculation on the subject. It was in the true spirit of philosophy that he so readily gave up an hypothesis, in which he could not but feel some interest, the moment he found it at variance with observation. He was sensible that nothing but the exact coincidence of the things compared could establish the conclusion he meant to deduce, or authorize him to proceed with the superstructure, for which it was to serve as the foundation.

It appears, that it was not till some years after this that his attention was called to the same subject, by a letter from Dr Hooke, proposing, as a question, To determine the line in which a body let fall from a height descends to the ground, taking into consideration the motion of the earth on its axis. This induced him to resume the subject of the moon's motion; and the measure of a degree by Norwood having now furnished more exact data, he found that his calculation gave the precise quantity for the moon's momentary deflection from the tangent of her orbit, which was deduced from astronomical observation. The moon, therefore, has a tendency to descend toward the earth from the same cause that a stone at its surface has; and if the descent of the stone in a second be diminished in the ratio of 1 to 3600, it will give the quantity by which the moon descends in a second, below the tangent to her orbit, and thus is obtained an experimental proof of the fact, that gravity decreases as the square of the distance increases. He had already found that the times of the planetary revolutions, supposing their orbits to be circular, led to the same conclusion; and he now proceeded, with a view to the solution of Hooke's problem, to inquire what their orbits must be, supposing the centripetal force to be inversely as the square of the distance, and the initial or projectile force to be any whatsoever. On this subject Pemberton says, he composed (as he calls it) a dozen propositions, which probably were the same with those in the beginning of the *Principia*,—such as the description of equal areas in equal times, about the centre of force, and the ellipticity of the orbits described under the influence of a centripetal force that varied inversely as the square of the distances.

What seems very difficult to be explained is, that, after having made trial of his strength, and of the power of the instruments of investigation which he was now in possession of, and had entered by means of them on the noblest and most magnificent field of investigation that was ever yet opened to any of the human race, he again desisted from the pursuit, so that it was not till several years afterwards that the conversation of Dr Halley, who made him a visit at Cambridge, induced him to resume and extend his researches.

He then found, that the three great facts in astronomy, which form the laws of Kepler, gave the most complete evidence to the system of gravitation. The *first* of them, the proportionality of the areas described by the radius vector to the times in which they are described, is the peculiar character of the motions produced by an original impulse impressed on a body, combined with a centripetal force continually urging it to a given centre. The *second* law, that the planets describe ellipses, having the sun in one of the foci, common to them all, coincides with this proposition, that a body under the influence of a centripetal force, varying as the square of the distance inversely, and having any projectile force whatever originally impressed on it, must describe a conic section having one focus in the centre of force, which section, if the projectile force does not exceed a certain limit, will become an ellipse. The *third* law, that the squares of the periodic times are as the cubes of the distances, is a property which belongs to the bodies describing elliptic orbits under the conditions just stated. Thus the three great truths to which the astronomy of the planets had been reduced by Kepler, were all explained in the most satisfactory manner, by the supposition that the planets gravitate to the sun with a force which varies in the inverse ratio of the square of the distances. It added much to this evidence, that the observations of Cassini had proved the same laws to prevail among the satellites of Jupiter.

But did the principle which appeared thus to unite the great bodies of the universe act only on those bodies? Did it reside merely in their centres, or was it a force common to all the particles of matter? Was it a fact that every particle of matter had a tendency to unite with every other? Or was that tendency directed only to particular centres? It could hardly be doubted that the tendency was common to all the particles of matter. The centres of the great bodies had no properties as mathematical points, they had none but what they derived from the material particles distributed around them. But the question admitted of being brought to a better test than that of such general reasoning as the preceding. The bodies between which this tendency had been observed to take place were all round bodies, and either spherical or nearly so, but whether great or small, they seemed to gravitate toward one another according to the same law. The planets gravitated to the sun, the moon to the earth, the satellites of Jupiter toward Jupiter; and gravity, in all these instances, varied inversely as the squares of the distances. Were the bodies ever so small—were they mere particles—provided only they were round, it was therefore safe to infer, that they would tend to unite with forces inversely as the squares of the distances. It was probable, then, that gravity was the mutual tendency of all the particles of matter toward one another; but this could not be concluded with certainty, till it was found, whether great spherical bodies composed of particles gravitating according to this law, would themselves gravitate according to the same. Perhaps no man of that age but Newton himself was fit to undertake the solution of this problem. His analysis, either in the form of fluxions or in that of prime and ultimate ratios, was able to reduce it to the quadrature of curves, and he then found, no doubt infinitely to his satisfaction, that the law was the same for the sphere as for the particles which compose it; that the gravitation was directed to the centre of the sphere, and was as the quantity of matter contained in it, divided by the square of the distance from its centre. Thus a complete expression was obtained for the law of gravity, involving both the conditions on which it must depend, the quantity of matter in the gravitating bodies, and the distance at which the bodies were placed. There could be no doubt that this tendency was always mutual, as there appeared nowhere any exception to the rule that action and reaction are equal; so that if a stone gravitated to the earth, the earth gravitated equally to the stone; that is to say, that the two bodies tended to approach one another with velocities which were inversely as their quantities of matter.* There appeared to be no limit to the distance to which this action reached; it was a force that united all the parts of matter to one another, and if it appeared to be particularly directed to certain points, such as the centres of the sun or of the planets, it was only on account of the quantity of matter collected and distributed uniformly round those points, through which, therefore, the force resulting from

---

* If M and M' are the masses of two spheres, and x the distance of their centres, $\frac{M+M'}{x^2}$ is the accelerating force with which they tend to unite; but the velocity of the approach of M will be $\frac{M'}{x^2}$, and of M', $\frac{M}{x^2}$.

the composition of all those elements must pass either accurately or nearly.

A remarkable inference was deduced from this view of the planetary motions, giving a deep in sight into the constitution of our system in a matter that seems the most recondite, and the furthest beyond the sphere which necessarily circumscribes human knowledge. The quantity of matter, and even the density of the planets, was determined. We have seen how Newton compared the intensity of gravitation at the surface of the earth, with its intensity at the moon, and by a computation somewhat similar, he compared the intensity of the earth's gravitation to the sun, with the moon's gravitation to the earth, each being measured by the contemporaneous and momentary deflexion from a tangent to the small arch of its orbit. A more detailed investigation showed that the intensity of the central force in different orbits, is as the mean distance divided by the square of the periodic time; and the same intensity being also as the quantities of matter divided by the squares of the distances, it follows, that these two quotients are equal to one another, and that, therefore, the quantities of matter are as the mean distances divided by the squares of the periodic times. Supposing, therefore, in the instance just mentioned, that the ratio of the mean distance of the sun from the earth to the mean distance of the moon from the earth is given, (which it is from astronomical observation;) as the ratio of their periodic lines is also known, the ratio of the quantity of matter in the sun to the quantity of matter in the earth, of consequence is found, and the same holds good for all the planets which have satellites moving round them. Nothing certainly can be more unexpected than that the quantities of matter in bodies so remote, should admit of being compared with one another, and with the earth. Hence also their mean densities, or mean specific gravities, became known. For from their distances and the angles they subtended, both known from observation, their magnitudes or cubical contents were easily inferred, and the densities of all bodies are, as their quantities of matter, divided by their magnitude. The Principia Philosophiæ Naturalis, which contained all these discoveries, and established the principle of universal gravitation, was given to the world in 1687, an æra, on that account, for ever memorable in the history of human knowledge.

The principle of gravity which was thus fully established, and its greatest and most extensive consequences deduced, was not now mentioned for the first time, though for the first time its existence as a fact was ascertained, and the law it observes was discovered. Besides some curious references to weight and gravity, contained in the writings of the ancients, we find something more precise concerning it in the writings of Copernicus, Kepler, and Hooke.

Anaxagoras is said to have held that "the heavens are kept in their place by the rapidity of their revolution, and would fall down if that rapidity were to cease." *

Plutarch, in like manner, says, the moon is kept from falling by the rapidity of her motion, just as a stone whirled round in a sling is prevented from falling to the ground. †

Lucretius, reasoning probably after Democritus, holds, that the atoms would all, from their gravity, have long since united in the centre of the universe, if the universe were not infinite so as to have no centre. ‡

An observation of Pythagoras, supposed to refer to the doctrine of gravity, though in reality extremely vague, has been abundantly commented on by Gregory and Maclaurin. A musical string, said that philosopher, gives the same sound with another of twice the length, if the latter be straitened by four times the weight that straitens the former; and the gravity of a planet is four times that of another which is at twice the distance. These are the most precise notices, as far as I know, that exist in the writings of the ancients concerning gravity as a force acting on terrestrial bodies, or as extending even to those that are more

* Cœlum omne vehementi circuitu constare, alias remissione lapsurum. (Diog. Laert. in Anax. Lib. II. Sect. 12.)

† De facie in Orbe Lunæ. ‡ Lib. I. v. 983.

distant. They are the reveries of ingenious men who had no steady principles deduced from experience and observation to direct their inquiries; and who, even when in their conjectures they hit on the truth, could hardly distinguish it from error.

Copernicus, as might be expected, is considerably more precise. " I do not think," says he, " that gravity is anything but a natural *appetency* of the parts (of the earth) given by the providence of the Supreme Being, that, by uniting together, they may assume the form of a globe. It is probable, that this same affection belongs to the sun, the moon, and the fixed stars, which all are of a round form." *

The power which Copernicus here speaks of has nothing to do, in his opinion, with the revolutions of the earth or the planets in their different orbits. It is merely intended as an explanation of their globular forms, and the consideration that does the author most credit is, that of supposing the force to belong, not to the centre, but to all the parts of the earth.

Kepler, in his immortal work on the Motions of Mars, treats of gravity as a force acting naturally from planet to planet, and particularly from the earth to the moon. " If the moon and the earth were not retained by some animal or other equivalent force each in its orbit, the earth would ascend to the moon by a 54th part of the interval between

---

* Revolutionum, Lib. I. Cap. IX. p. 17.

them, while the moon moved over the remaining 53 parts, that is, supposing them both of the same density." * This passage is curious, as displaying a singular mixture of knowledge and error on the subject of the planetary motions. The tendency of the earth and moon being mutual, and producing equal quantities of motion in those bodies, bespeaks an accurate knowledge of the nature of that tendency, and of the equality, at least in this instance, between action and reaction. Then, again, the idea of an animal force or some other equally unintelligible power being necessary to carry on the circular motion, and to prevent the bodies from moving directly toward each other, is very strange ; considering that Kepler knew the inertia of matter, and ought, therefore, to have understood the nature of centrifugal force, and its power to counteract the mutual gravitations of the two bodies. In this respect, the great astronomer who was laying the foundation of all that is known of the heavens, was not so far advanced as Anaxagoras and Plutarch ; —so slow and unequal are the steps by which science advances to perfection. The mutual gravity of the earth and moon is not supposed by Kepler to have any concern in the production of their circular motions ; yet he holds the tides to be produced by

* On that supposition their quantities of matter would be as their bulks, or as 1 to 53.

the gravitation of the waters of the sea toward the moon. *

The length to which Galileo advanced in this direction, and the point at which he stopped, are no less curious to be remarked. Though so well acquainted with the nature of gravity on the earth's surface,—the object of so many of his researches and discoveries,—and though he conceived it to exist in all the planets, nay, in all the celestial bodies, and to be the cause of their round figure, he did not believe it to be a power that extended from one of those bodies to another. He seems to have thought that gravity was a principle which regulated the domestic economy of each particular body, but had nothing to do with their external relations; so that he censured Kepler for supposing, that the phenomena of the tides are produced by the gravitation of the waters of the ocean to the moon. †

Hooke did not stop short in the same unaccountable manner, but made a nearer approach to the truth than any one had yet done. In his attempt to prove the motion of the earth, published in 1674, he lays it down as the principle on which the celestial motions are to be explained, that the heavenly bodies have an attraction or gravitation toward their own centres, which extends to other bo-

* Astronomia Stellæ Martis. Introd. Parag. 8.

† Dial. 4to. Tom. IV. p. 325, Ediz. di Padova.

dies within the spnere of their activity ; and that all bodies would move in straight lines, if some force like this did not act on them continually, and compel them to describe circles, ellipses, or other curve lines. The force of gravity, also, he considered as greatest nearest the body, though the law of its variation he could not determine. These are great advances ;—though, from his mention of the sphere of activity, from his considering the force as residing in the centre, and from his ignorance of the law which it observed, it is evident, that beside great vagueness, there was much error in his notions about gravity. Hooke, however, whose candour and uprightness bore no proportion to the strength of his understanding, was disingenuous enough, when Newton had determined that law, to lay claim himself to the discovery.

This is the farthest advance that the knowledge of the cause of the celestial motions had made before the investigations of Newton ; it is the precise point at which this knowledge had stopped ; having met with a resistance which required a mathematician armed with all the powers of the new analysis to overcome. The doctrine of gravity was yet no more than a conjecture, of the truth or falsehood of which the measurements and reasonings of geometry could alone determine.

Thus, then, we are enabled accurately to perceive in what Newton's discovery consisted. It was in giving the evidence of demonstration to a principle which a few sagacious men had been sufficiently sharp-sighted to see obscurely or inaccurately, and to propose as a mere conjecture. In the history of human knowledge, there is hardly any discovery to which some gradual approaches had not been made before it was completely brought to light. To have found out the means of giving certainty to the thing asserted, or of disproving it entirely; and, when the reality of the principle was found out, to measure its quantity, to ascertain its laws, and to trace their consequences with mathematical precision,—in this consists the great difficulty and the great merit of such a discovery as that which is now before us. In this Newton had no competitor: envy was forced to acknowledge that he had no rival, and consoled itself with supposing that he had no judge.

Of all the physical principles that have yet been made known, there is none so fruitful in consequences as that of gravitation; but the same skill that had directed Newton to the discovery was necessary to enable him to trace its consequences.

The mutual gravitation of all bodies being admitted, it was evident, that while the planets were describing their orbits round the greatest and most powerful body in the system, they must mutually attract one another, and thence, in their revolutions, some irregularities, some deviations from the description of equal areas in equal times, and from the laws of the elliptic motion might be expected. Such irregularities, however, had not been observed at that time in the motion of any of the planets, except the moon, where some of them were so conspicuous as to have been known to Hipparchus and Ptolemy. Newton, therefore, was very naturally led to inquire what the different forces were, which, according to the laws just established, could produce irregularities in the case of the moon's motion. Beside the force of the earth, or rather of the mutual gravitation of the moon and earth, the moon must be acted on by the sun; and the same force which was sufficient to bend the orbit of the earth into an ellipse, could not but have a sensible effect on the orbit of the moon. Here Newton immediately observed, that it is not the whole of the force which the sun exerts on the moon that disturbs her motion round the earth, but only the difference between the force just mentioned, and that which the sun exerts on the earth,—for it is only that difference that affects the relative positions of the two bodies. To have exact measures of the disturbing forces, he supposed the entire force of the sun on the moon to be resolved into two, of which one always passed through the centre of the earth, and the other was always parallel to the line joining the sun and earth,—consequently, to the direction of the force of the sun on the earth.

The former of these forces being directed to the centre of the earth, did not prevent the moon from describing equal areas in equal times round the earth. The effect of it on the whole, however, he showed to be, to diminish the gravity of the moon to the earth by about one 358th part, and to increase her mean distance in the same proportion, and her angular motion by about a 179th.

From the moon thus gravitating to the centre of the earth, not by a force that is altogether inversely as the square of the distance, but by such a force diminished by a small part that varies simply as the distance, it was found, from a very subtle investigation, that the dimensions of the elliptic orbit would not be sensibly changed, but that the orbit itself would be rendered moveable, its longer axis having an angular and progressive motion, by which it advanced over a certain are during each revolution of the moon. This afforded an explanation of the motion of the apsides of the lunar orbit which had been observed to go forward at the rate of 3° 4′, nearly, during the time of the moon's revolution, in respect of the fixed stars.

This was a new proof of the reality of the principle of gravitation, which, however, was rendered less conclusive by the consideration that the exact quantity of the motion of the apsides observed, did not come out from the diminution of the moon's gravity as above assigned. There was a sort of cloud, therefore, which hung over this point of the lunar theory, to dissipate which, required higher improvements in the calculus than it was given to the inventor himself to accomplish. It was not so with respect to another motion to which the plane of the lunar orbit is subject, a phenomenon which had been long known in consequence of its influence on the eclipses of the sun and moon. This was the retrogradation of the line of nodes, amounting to 3′ 10″ every day. Newton showed that the second of the forces into which the solar action is resolved being exerted, not in the plane of the moon's orbit, but in that of the ecliptic, inclined to the former at an angle somewhat greater than five degrees, its effect must be to draw down the moon to the plane of the ecliptic sooner than it would otherwise arrive at it; in consequence of which, the intersection of the two planes would approach, as it were, toward the moon, or move in a direction opposite to that of the moon's motion, or become retrograde. From the quantity of the solar force, and the inclination of the moon's orbit, Newton determined the mean quantity of this retrogradation, as well as the irregularities to which it is subject, and found both to agree very accurately with observation.

Another of the lunar inequalities,—that discovered by Tycho, and called by him the *Variation*, which consists in the alternate acceleration and retardation of the moon in each quarter of her revolution, was accurately determined from theory, such as it is found by observation; and the same is true as to the annual equation, which had been long confounded with the equation of time. With regard to the other inequalities, it does not appear that Newton attempted an exact determination of them, but satisfied himself with this general truth, that the principle of the sun's disturbing force led to the supposition of inequalities of the same kind with those actually observed, though whether of the same exact quantity it must be difficult to determine. It was reserved, indeed, for a more perfect state of the calculus to explain the whole of those irregularities, and to deduce their precise value from the theory of gravity. Theory has led to the knowledge of many inequalities, which observation alone would have been unable to discover.

While Newton was thus so successfully occupied in tracing the action of gravity among those distant bodies, he did not, it may be supposed, neglect the consideration of its effects on the objects which are nearer us, and particularly on the Figure of the Earth. We have seen that, even with the limited views and imperfect information which Copernicus possessed on this subject, he ascribed the round figure of the earth and of the planets to the force of gravity residing in the particles of these bodies.

Newton, on the other hand, perceived that, in the earth, another force was combined with gravity, and that the figure resulting from that combination could not be exactly spherical. The diurnal revolution of the earth, he knew, must produce a centrifugal force, which would act most powerfully on the parts most distant from the axis. The amount of this centrifugal force is greatest at the equator, and being measured by the momentary recess of any point from the tangent, which was known from the earth's rotation, it could be compared with the force of gravity at the same place, measured in like manner by the descent of a heavy body in the first moment of its fall. When Newton made this comparison, he found that the centrifugal force at the equator is the 289th part of gravity, diminishing continually as the cosine of the latitude, on going from thence toward the poles, where it ceases altogether. From the combination of this force, though small, with the force of gravity, it follows, that the line in which bodies actually gravitate, or the plumb-line, cannot tend exactly to the earth's centre, and that a true horizontal line, such as is drawn by levelling, if continued from either pole, in the plane of a meridian all round the earth, would not be a circle but an ellipse, having its greatest axis in the plane of the equator, and its least in the direction of the axis of the earth's rotation. Now, the surface of the ocean itself actually traces this level as it extends from the equator to either pole. The terraqueous mass which we call the globe must therefore be what geometers call an oblate spheroid, or a solid generated by the revolution of the elliptic meridian about its shorter axis.

In order to determine the proportion of the axes of this spheroid, a problem, it will readily be believed, of no ordinary difficulty, Newton conceived, that if the waters at the pole and at the equator were to communicate by a canal through the interior of the earth, one branch reaching from the pole to the centre, and the other at right angles to it, from the centre to the circumference of the equator, the water in this canal must be *in equilibrio*, or the weight of fluid in the one branch just equal to that in the other. Including, then, the consideration of the centrifugal force which acted on one of the branches but not on the other, and considering, too, that the figure of the mass being no longer a sphere, the attraction must not be supposed to be directed to the centre, but must be considered as the result of the action of all the particles of the spheroid on the fluid in the canals; by a very subtle process of reasoning, Newton found that the longer of the two canals must be to the shorter as 230 to 229. This, therefore, is the ratio of the radius of the equator to the polar semiaxis, their difference amounting, according to the dimensions then assigned to the earth, to about $17\frac{1}{10}$ English miles. In this investigation, the earth is understood to be homogeneous, or everywhere of the same density.

It is very remarkable, that though the ingenious and profound reasoning on which this conclusion rests is not entirely above objection, and assumes some things without sufficient proof, yet, when these defects were corrected in the new investigations of Maclaurin and Clairaut, the conclusion, supposing the earth homogeneous, remained exactly the same. The sagacity of Newton, like the *Genius* of Socrates, seemed sometimes to inspire him with wisdom from an invisible source. By a profound study of nature, her laws, her analogies, and her resources, he seems to have acquired the same sort of *tact* or *feeling* in matters of science, that experienced engineers and other artists sometimes acquire in matters of practice, by which they are often directed right, when they can scarcely describe in words the principle on which they proceed.

From the figure of the earth thus determined, he showed that the intensity of gravity at any point of the surface, is inversely as the distance of that point from the centre ; and its increase, therefore, on going from the equator to the poles, is as the square of the sine of the latitude, the same ratio in which the degrees of the meridian increase. * As

* Princip. Lib. III. Prop. 20.

the intensity of gravity diminished on going from the poles to the equator, or from the higher to the lower latitudes, it followed, that a pendulum of a given length would vibrate slower when carried from Europe into the torrid zone. The observations of the two French astronomers, Varin and De Hayes, made at Cayenne and Martinique, had already confirmed this conclusion.

The problem which Newton had thus resolved enabled him to resolve one of still greater difficulty. The precession, that is, the retrogradation of the equinoctial points, had been long known to astronomers; its rate had been measured by a comparison of ancient and modern observations, and found to amount nearly to 50″ annually, so as to complete an entire revolution of the heavens in 25,920 years. Nothing seemed more difficult to explain than this phenomenon, and no idea of assigning a physical or mechanical cause for it had yet occurred, I believe, to the boldest and most theoretical astronomer. The honour of assigning the true cause was reserved for the most cautious of philosophers. He was directed to this by a certain analogy observed between the precession of the equinoxes and the retrogradation of the moon's nodes, a phenomenon to which his calculus had been already successfully applied. The spheroidal shell or ring of matter which surrounds the earth, as we have just seen, in the direction of the equator, being one half above the plane of the ecliptic and the other half below, is subjected to the action of the solar force, the tendency of which is to make this ring turn on the line of its intersection with the ecliptic, so as ultimately to coincide with the plane of that circle. This, accordingly, would have happened long since, if the earth had not revolved on its axis. The effect of the rotation of the spheroidal ring from west to east, at the same time that it is drawn down toward the plane of the ecliptic, is to preserve the inclination of these two planes unchanged, but to make their intersection move in a direction opposite to that of the diurnal rotation, that is, from east to west, or contrary to the order of the signs.

The calculus in its result justified this general conclusion ; 10″ appeared the part of the effect due to the moon's attraction, 40″ to the attraction of the sun ; and I know not if there be any thing respecting the constitution of our system, in which this great philosopher gave a stronger proof of his sagacity and penetration, than in the explanation of this phenomenon. The truth, however, is, that his data for resolving the problem were in some degree imperfect, all the circumstances were not included, and some were erroneously applied, yet the great principle and scope of the solution were right, and the approximation very near to the truth. " Il a été bien servi par son génie," says the eloquent and judicious historian of astronomy; "l'inspiration de cette faculté divine lui a fait apercevoir des déterminations, qui n'étoient pas encore accessibles; soit qu'il eût des preuves qu'il a supprimées, *soit qu'il eût dans l'esprit un sorte d'estime, une espèce de balance pour approuver certaines vérités, en pesant les verités prochaines, et jugeant les unes par les autres.*" *

It was reserved for a more advanced condition of the new analysis, to give to the solution of this problem all the accuracy of which it is susceptible. It is a part, and a distinguishing part, of the glory of this system, that it was susceptible of more perfection than it received from the hands of the author; and that the century and a half which has nearly elapsed since the first discovery of it has been continually adding to its perfection. This character belongs to a system which has truth and nature for its basis, and had not been exhibited in any of the physical theories that had yet appeared in the world. The philosophy of Plato and Aristotle were never more perfect than when they came from the hands of their respective authors, and a legion of commentators, with all their efforts, did nothing but run round perpetually in the same circle. Even Descartes, though he had recourse to physical principles, and tried to fix his system

* Bailly, Hist. de l'Astron. Mod. Tom. II. livre xii. § 28.

on a firmer basis than the mere abstractions of the mind, left behind him a work which not only could not be improved, but was such, that every addition attempted to be made destroyed the equilibrium of the mass, and pulled away the part to which it was intended that it should be attached. The philosophy of Newton has proved susceptible of continual improvement; its theories have explained facts quite unknown to the author of it; and the exertions of La Grange and La Place, at the distance of an hundred years, have perfected a work which it was not for any of the human race to begin and to complete.

Newton next turned his attention to the phenomena of the Tides, the dependence of which on the moon, and in part also on the sun, was sufficiently obvious even from common observation. That the moon is the prime ruler of the tide, is evident from the fact, that the high water, at any given place, occurs always nearly at the moment when the moon is on the same meridian, and that the retardation of the tide from day to day, is the same with the retardation of the moon in her diurnal revolution. That the sun is also concerned in the production of the tides is evident from this, that the highest tides happen when the sun, the moon, and the earth, are all three in the same straight line; and that the lowest, or neap tides, happen when the lines drawn from the sun and moon to the earth make right angles with one another. The eye of Newton, accustomed to generalize and to penetrate beyond the surface of things, saw that the waters of the sea revolving with the earth, are nearly in the condition of a satellite revolving about its primary; and are liable to the same kind of disturbance from the attraction of a third body. The fact in the history of the tides which seems most difficult to be explained, received, on this supposition, a very easy solution. It is known, that high water always takes place in the hemisphere where the moon is, and in the opposite hemisphere where the moon is not, nearly at the same time. This seems, at first sight, very unlike an effect of the moon's attraction; for, though the water in the hemisphere where the moon is, and which, therefore, is nearest the moon, may be drawn up toward that body, the same ought not to happen in the opposite hemisphere, where the earth's surface is most distant from the moon. But if the action of the moon disturb the equilibrium of the ocean, just as the action of one planet disturbs the motion of a satellite moving round another, it is exactly what might be expected. It had been shown, that the moon, in conjunction with the sun, has her gravitation to the earth diminished, and when in opposition to the sun, has it diminished very nearly by the same quantity. The reason is, that at the conjunction, or the new moon, the moon is drawn to the sun more than the earth is; and that, at the opposition, or full moon, the earth is drawn toward the sun more than the moon nearly by the same quantity; the relative motion of the two bodies is therefore affected the same way in both cases, and the gravity of the moon to the earth, or her tendency to descend toward it, is in both cases lessened.

It is plain, that the action of the moon on the waters of the ocean must be regulated by the same principle. In the hemisphere where the moon is, the water is more drawn toward the moon than the mass of the earth is, and its gravity being lessened, the columns toward the middle of the hemisphere lengthen, in consequence of the pressure of the columns which are at a distance from the middle point, of which the weight is less diminished, and towards the horizon must even be increased. In the opposite hemisphere, again, the mass of the earth is more drawn to the moon than the waters of that hemisphere, and their relative tendencies are changed in the same direction, and nearly by the same quantity. If the action of the moon on all the parts of the earth, both sea and land, were the same, no tide whatever would be produced.

Thus, the same analysis of the force of gravity which explained the inequalities of the moon, was shown by Newton to explain those inequalities in the elevation of the waters of the ocean to which we give the name of tides. On the principle also explained in this analysis, it is, that the attraction of the sun and moon conspire to elevate the waters of the ocean whether these luminaries be in opposition or conjunction. In both cases the solar and lunar tides are added together, and the tide actually observed is their sum. At the quadratures, or the first and third quarters, these two tides are opposed to one another, the high water of the lunar tide coinciding with the low water of the solar, and conversely, so that the tide actually observed is the difference of the two.

The other phenomena of the tides were explained in a manner no less satisfactory, and it only remained to inquire, Whether the quantity of the solar and lunar forces were adequate to the effect thus ascribed to them? The lunar force there were yet no data for measuring, but a measure of the solar force, as it acts on the moon, had been obtained, and it had been shown that in its mean quantity it amounted to $\frac{1}{178}$ of the force which retains the moon in her orbit. This last is $\frac{1}{3600}$ of the force of gravity at the earth's surface, and, therefore, the force with which the sun disturbs the moon's motion is $\frac{1}{178} \times \frac{1}{3600}$ of gravity at the earth's surface. This is the solar disturbing force on the moon when distant sixty semidiameters from the earth's centre, but on a body only one semidiameter distant from that centre, that is, on the water of the ocean, the disturbing force would be sixty times less, and thus is found to be no more than $\frac{1}{38448000}$ of gravity at the earth's surface.

Now, this being the mean force of the sun, is that by which he acts on the waters, 90 degrees distant from the point to which he is vertical, where it is added to the force of gravity, and tends to increase the weight and lower the level of the waters. At the point where the sun is vertical, the force to raise the water is about double of this, and, therefore, the whole force tending to raise the level of the high, above that of the low water, is three times the preceding, or about the $\frac{1}{12816000}$ of gravity. Small as this force is, when it is applied to every particle of the ocean, it is capable of producing a sensible effect. The manner in which Newton estimates this effect can only be considered as affording an approximation to the truth. In treating of the figure of the earth, he had shown that the centrifugal force, amounting to $\frac{1}{289}$ of gravity, was able to raise the level of the ocean more than seventeen miles, or, more exactly, 85,472 French feet. Hence, making the effect proportional to the forces, the elevation of the waters produced by the solar force will come out 1.92 feet.

But, from the comparison of the neap and spring tides, that is, of the difference and the sum of the lunar and solar forces, it appears, that the force of the moon is to that of the sun as 4.48 to 1. As the solar force raises the tide 1.92 feet, the lunar will raise it 8.63 feet, so that the two together will produce a tide of $10\frac{1}{2}$ French feet,* which agrees not ill with what is observed in the open sea, at a distance from land.

The calculus of Newton stopped not here. From the force that the moon exerts on the waters of the ocean, he found the quantity of matter in the moon to that in the earth as 1 to 39.78, or, in round numbers, as 1 to 40. He also found the density of the moon to the density of the earth as 11 to 9.

Subsequent investigations, as we shall have occasion to remark, have shown that much was yet wanting to a complete theory of the tides; and that even after Maclaurin, Bernoulli, and Euler,† had added their efforts to those of Newton, there remained enough to give full employment to the calculus of Laplace. As an original deduction, and as a first approximation, that of which I have now given an account, will be for ever memorable.

The motion of Comets yet remained to be discussed. They had only lately been acknowledged to belong to the heavens, and to be placed beyond the region of the earth's atmosphere; but with regard to their motion, astronomers were not agreed.

* Newtoni, Prin. Lib. III. Prop. 36 ad 37.

† See the solutions of these three mathematicians in the Commentary of Le Seur and Jacquier on the Third Book of the Principia.

Kepler believed them to move in straight lines; Cassini thought they moved in the planes of great circles, but with little curvature. Hevelius had come much nearer the truth; he had shown the curvature of their paths to be different in different parts, and to be greatest when they were nearest the sun; and a parabola having its vertex in that point seemed to him to be the line in which the comet moved. Newton, convinced of the universality of the principle of gravitation, had no doubt that the orbit of the comet must be a conic section, having the sun in one of its foci, and might either be an ellipse, a parabola, or even an hyperbola, according to the relation between the force of projection and the force tending to the centre. As the eccentricity of the orbit on every supposition must be great, the portion of it that fell within our view could not differ much from a parabola, a circumstance which rendered the calculation of the comet's place, when the position of the orbit was once ascertained, more easy than in the case of the planets. Thus far theory proceeded, and observation must then determine with what degree of accuracy this theory represented the phenomena. From three observations of the comet, the position of the orbit could be determined, though the geometric problem was one of great difficulty. Newton gave a solution of it; and it was by this that his theory was to be brought to the test of experiment. If the orbit thus determined was not the true one, the places of the comet calculated on the supposition that it was, and that it described equal areas in equal times about the sun, could not agree with the places actually observed. Newton showed, by the example of the remarkable comet then visible, (1680,) that this agreement was as great as could reasonably be expected; thus adding another proof to the number of those already brought to support the principle of universal gravitation. The comets descend into our system from all different quarters in the heavens, and, therefore, the proofs that they afforded went to show, that the action of gravity was confined to no particular region of the heavens.

Thus far Newton proceeded in ascertaining the existence, and in tracing the effects, of the principle of gravitation, and had done so with a success of which there had been no instance in the history of human knowledge. At the same time that it was the most successful, it was the most difficult research that had yet been undertaken. The reasonings upward from the facts to the general principle, and again down from that principle to its effects, both required the application of a mathematical analysis which was but newly invented; and Newton had not only the difficulties of the investigation to encounter, but the instrument to invent, without which the investigation could not have been conducted. Every one who considers all this, will readily join in the sentiment with which Bailly closes a eulogy as just as it is eloquent. *Si, comme Platon a pensé, il existoit dans la nature une echelle d'êtres et de substances intelligentes jusqu'à l'Etre Suprême, l'espéce humaine, défendant ses droits, auroit une foule de grands hommes à presenter; mais Newton, suivi de ses vérités pures, montreroit le plus haut degré de force de l'esprit humain, et suffiroit seul pour lui assigner sa vrai place.* *

Though the creative power of genius was never more clearly evinced than in the discoveries of this great philosopher, yet the influence of circumstances, always extensive and irresistible in human affairs, can readily be traced. The condition of knowledge at the time when Newton appeared was favourable to great exertions; it was a moment when things might be said to be prepared for a revolution in the mathematical and physical sciences. The genius of Copernicus had unfolded the true system of the world; and Galileo had shown its excellence, and established it by arguments, the force of which were generally acknowledged. Kepler had done still more, having, by an admirable effort of generalisation, reduced the facts concerning the planetary motions to three general laws. Cassini's observations had also extended the third of

* Hist. de l'Astron. Mod. Tom. 11.

these laws to the satellites of Jupiter, showing that the squares of their periodic times were as the cubes of their distances from the centre of the body round which they revolved. The imaginary apparatus of cycles and epicycles,—the immobility of the earth,—the supposed essential distinction between celestial and terrestrial substances, those insuperable obstacles to real knowledge, which the prejudice of the ancients had established as physical truths, were entirely removed; and Bacon had taught the true laws of philosophising, and pointed out the genuine method of extracting knowledge from experiment and observation. The leading principles of mechanics were established; and it was no unimportant circumstance, that the *Vortices* of Descartes had exhausted one of the sources of error, most seducing on account of its simplicity.

All this had been done when the genius of Newton arose upon the earth. Never till now had there been set before any of the human race so brilliant a career to run, or so noble a prize to be obtained. In the progress of knowledge, a moment had arrived more favourable to the developement of talent than any other, either later or earlier, and in which it might produce the greatest possible effect. But, let it not be supposed, while I thus admit the influence of external circumstances on the exertions of intellectual power, that I am lessening the merit of this last, or taking any thing from the admiration that is due to it. I am, in truth, only distinguishing between what it is possible, and what it is impossible, for the human mind to effect. With all the aid that circumstances could give, it required the highest degree of intellectual power to accomplish what Newton performed. We have here a memorable, perhaps a singular instance, of the highest degree of intellectual power, united to the most favourable condition of things for its exertion. Though Newton's situation was more favourable than that of the men of science who had gone before him, it was not more so than that of those men who pursued the same objects at the same time with himself, placed in a situation equally favourable.

When one considers the splendour of Newton's discoveries, the beauty, the simplicity, and grandeur of the system they unfolded, and the demonstrative evidence by which that system was supported, one could hardly doubt, that, to be received, it required only to be made known, and that the establishment of the Newtonian philosophy all over Europe would very quickly have followed the publication of it. In drawing this conclusion, however, we should make much too small an allowance for the influence of received opinion, and the resistance that mere habit is able, for a time, to oppose to the strongest evidence. The Cartesian system of vortices had many followers in all the countries of Europe, and particularly in France. In the universities of England, though the Aristotelian physics had made an obstinate resistance, they had been supplanted by the Cartesian, which became firmly established about the time when their foundation began to be sapped by the general progress of science, and particularly by the discoveries of Newton. For more than thirty years after the publication of those discoveries, the system of vortices kept its ground, and a translation from the French into Latin of the Physics of Rohault, a work entirely Cartesian, continued at Cambridge to be the text for philosophical instruction. About the year 1718, a new and more elegant translation of the same book was published by Dr Samuel Clarke, with the addition of notes, in which that profound and ingenious writer explained the views of Newton on the principal objects of discussion, so that the notes contained *virtually* a refutation of the text; they did so, however, only virtually, all appearance of argument and controversy being carefully avoided. Whether this escaped the notice of the learned Doctors or not is uncertain, but the new translation, from its better Latinity, and the name of the editor, was readily admitted to all the academical honours which the old one had enjoyed. Thus, the stratagem of Dr Clarke completely succeeded; the tutor might prelect from the text, but the pupil would sometimes look into the notes, and error is never so sure of being exposed as when the truth is placed close to it, side by side, without any thing to alarm prejudice, or awaken from its lethargy the dread of innovation. Thus, therefore, the *Newtonian* philosophy first entered the University of Cambridge under the protection of the *Cartesian.* *

---

* The Universities of St Andrews and Edinburgh were, I believe, the first in Britain where the Newtonian philosophy was made the subject of the academical prelections. For this distinction they are indebted to James and David Gregory, the first in some respects the rival, but both the friends of Newton. Whiston bewails, in the anguish of his heart, the difference in this respect between those universities and his own. David Gregory taught in Edinburgh for several years prior to 1690, when he removed to Oxford; and Whiston says, "He had already caused several of his scholars to keep acts, as we call them, upon several branches of the Newtonian philosophy, while we at Cambridge (poor wretches) were ignominiously studying the fictitious hypotheses of the Cartesian." (Whiston's Memoirs of his own Life.) I do not, however, mean to say, that from this date the Cartesian philosophy was expelled from those universities; the Physics of Rohault were still in use as a text, at least occasionally, to a much later period than this, and a great deal, no doubt, depended on the character of the individual professors. Keil introduced the Newtonian philosophy in his lectures at Oxford in 1697; but the instructions of the tutors, which constitute the real and efficient system of the university, were not cast in that mould till long afterwards. The publication of S'Gravesende's Elements

If such were the obstacles to its progress that the new philosophy experienced in a country that was proud of having given birth to its author, we must expect it to advance very slowly indeed among foreign nations. In France, we find the first astronomers and mathematicians, such men as Cassini and Maraldi, quite unacquainted with it, and employed in calculating the paths of the comets they were observing, on hypotheses the most unfounded and imaginary; long after Halley, following the principles of Newton, had computed tables from which the motions of all the comets that ever had appeared, or ever could appear, might be easily deduced. Fontenelle with great talents and enlarged views, and, as one may say, officially informed of the progress of science all over Europe, continued a Cartesian to the end of his days. Mairan in his youth was a zealous defender of the vortices, though he became afterwards one of the most strenuous supporters of the doctrine of gravitation.

A Memoir of the Chevalier Louville, among those of the Academy of Sciences for 1720, is the first in that collection, and, I believe, the first published in France, where the elliptic motion of the planets is supposed to be produced by the combination of two forces, one projectile and the other cen-

proves that the Newtonian philosophy was taught in the Dutch universities before the date of 1720.

tripetal. Maupertuis soon after went much farther; in his elegant and philosophic treatise, Figure des Astres, published about 1730, he not only admitted the existence of attraction as a fact, but even defended it, when considered as an universal property of body, against the reproach of being a metaphysical absurdity. These were considerable advances, but they were made slowly; and it was true, as Voltaire afterwards remarked, that though the author of the Principia survived the publication of that great work nearly forty years, he had not, at the time of his death, twenty followers out of England.

We should do wrong, however, to attribute this slow conversion of the philosophic world entirely to prejudice, inertness, or apathy. The evidence of the Newtonian philosophy was of a nature to require time in order to make an impression. It implied an application of mathematical reasoning which was often difficult; the doctrine of prime and ultimate ratios was new to most readers, and could be familiar only to those who had studied the infinitesimal analysis.

The principle of gravitation itself was considered as difficult to be admitted. When presented, indeed, as a mere fact, like the weight of bodies at the earth's surface, or their tendency to fall to the ground, it was free from objection; and it was in this light only that Newton wished it to be considered.* But though this appears to be the sound and philosophical view of the subject, there has always appeared a strong desire in those who speculated concerning gravitation to go farther, and to inquire into the cause of what, as a mere fact, they were sufficiently disposed to admit. If you said that you had no explanation to give, and was only desirous of having the fact admitted, they alleged, that this was an unsatisfactory proceeding, —that it was admitting the doctrine of *occult causes*,—that it amounted to the assertion, that bodies acted in places where they were not,—a proposition that, metaphysically considered, was undoubtedly absurd. The desire to explain gravitation is indeed so natural, that Newton himself felt its force, and has thrown out, at the end of his Optics, some curious conjectures concerning this general affection of body, and the nature of that elastic ether to which he thought that it was perhaps to be ascribed. "Is not this medium (the ether) much rarer within the dense bodies of the sun, stars, and planets, than in the empty celestial

* "*Vocem attractionis hic generaliter usurpo pro corporum conatu quocunque accedendi ad se invicem; sive conatus iste fiat ab actione corporum se mutuo petentium, vel per spiritus emissos se mutuo agitantium; sive is ab actione ætheris, aut aeris medii cujuscunque, corporei vel incorporei oritur, corpora innatantia in se invicem utcunque impellentes.*" Principia Math. Lib. 1. Schol. ad finem. Prop. 69.

spaces between them? And, in passing from them to great distances, does it not grow denser and denser perpetually, and thereby cause the gravity of those great bodies to one another, every body endeavouring to go from the denser parts of the medium to the rarer?" *

Notwithstanding the highest respect for the author of these conjectures, I cannot find any thing like a satisfactory explanation of gravity in the existence of this elastic ether. It is very true that an elastic fluid, of which the density followed the inverse ratio of the distance from a given point, would urge the bodies immersed in it, and impervious to it, toward that point with forces inversely as the squares of the distances from it; but what could maintain an elastic fluid in this condition, or with its density varying according to this law, is a thing as inexplicable as the gravity which it was meant to explain. The nature of an elastic fluid must be, in the absence of all inequality of pressure, to become every where of the same density. If the causes that produce so marked and so general a deviation from this rule be not assigned, we can only be said to have substituted one difficulty for another.

A different view of the matter was taken by some of the disciples and friends of Newton, but which

---

* Optics, Query 21, at the end of the Third Book.

certainly did not lead to any thing more satisfactory. That philosopher himself had always expressed his decided opinion * that gravity could not be consi-

---

* The passages quoted sufficiently prove that Newton did not consider gravity as a property inherent in matter. The following passage, in one of his Letters to Dr Bentley, is still more explicit: "It is inconceivable that inanimate brute matter should, without the mediation of something else, which is not material, operate upon and affect other matter without mutual contact; as it must do, if gravitation, in the sense of Epicurus, be essential or inherent in it. That gravity should be innate, inherent, and essential to matter, so that one body may act on another, at a distance, through a vacuum, without the mediation of any thing else, by and through which their action and force may be conveyed from one to another, is, to me, so great an absurdity, that I believe no man who, in philosophical matters, has a competent faculty of thinking can ever fall into it." (*Newtoni Opera*, Tom. IV. Horsley's edit. p. 438.) On this passage I cannot help remarking, that it is not quite clear in what manner the interposition of a material substance can convey the action of distant bodies to one another. In the case of percussion or pressure, this is indeed very intelligible, but it is by no means so in the case of attraction. If two particles of matter, at opposite extremities of the diameter of the earth, attract one another, this effect is just as little intelligible, and the *modus agendi* is just as mysterious, on the supposition that the whole globe of the earth is interposed, as on that of nothing whatever being interposed, or of a complete vacuum existing between them. It is not enough that each particle attracts that in contact with it; it must attract the particles that are distant, and the intervention of particles between hem does not render this at all more intelligible.

dered as a property of matter; but Mr Cotes, in the preface to the second edition of the Principia, maintains, that gravity is a property which we have the same right to ascribe to matter, that we have to ascribe to it extension, impenetrability, or any other property. This is said to have been inserted without the knowledge of Newton,—a freedom which it is difficult to conceive that any man could use with the author of the Principia. However that be, it is certain that these difficulties have been always felt, and had their share in retarding the progress of the philosophy to which they seemed to be inseparably attached.

There were other arguments of a less abstruse nature, and more immediately connected with experiment, which, for a time, resisted the progress of the Newtonian philosophy, though they contributed, in the end, very materially to its advancement. Nothing, indeed, is so hostile to the interests of truth, as facts inaccurately observed; of which we have a remarkable example in the measurement of an arch of the meridian across France, from Amiens to Perpignan, though so large as to comprehend about seven degrees, and though executed by Cassini, one of the first astronomers in Europe. According to that measurement, the degrees seemed to diminish on going from south to north, each being less by about an 800th part than that which immediately preceded it toward the south. From this result, which is entirely erroneous, the conclusion first deduced was correct, the error in the reasoning, by a very singular coincidence, having corrected the error in the *data* from which it was deduced. Fontenelle argued that, as the degrees diminished in length on going toward the poles, the meridian must be less than the circumference of the equator, and the earth, of course, swelled out in the plane of that circle, agreeably to the facts that had been observed concerning the retardation of the pendulum when carried to the south. This, however, was the direct contrary of the conclusion which ought to have been drawn, as was soon perceived by Cassini and by Fontenelle himself. The degrees growing less as they approached the pole, was an indication of the curvature growing greater, or of the longer axis of the meridian being the line that passed through the poles, and that coincided with the axis of the earth. The figure of the earth must, therefore, be that of an oblong spheroid, or one formed by the revolution of an ellipsis about its longer axis. This conclusion seemed to be strengthened by the prolongation of the meridian from Amiens northward to Dunkirk in 1713, as the same diminution was observed; the medium length of the degree between Paris and Dunkirk being 56970 toises, no less than 137 less than the mean of the degrees toward the south.* All this seemed quite inconsistent with the observations on the pendulum, as well as with the conclusions which Newton had deduced from the theory of gravity. The Academy of Sciences was thus greatly perplexed, and uncertain to what side to incline. In these circumstances, J. Cassini, whose errors were the cause of all the difficulty, had the merit of suggesting the only means by which the question concerning the figure of the earth was likely to receive a satisfactory solution,—the measurement of two degrees, the one under the equator, and the other as near to the pole as the nature of the thing would admit. But it was not till considerably beyond the limits of the period of which I am now treating, that these measures were executed; and that the increase of the degrees toward the poles, or the oblateness of the earth's figure, was completely ascertained. Cassini, on resuming his own operations, discovered, and candidly acknowledged, the errors in his first measurement; and thus the objections which had arisen in this quarter against the theory of gravity became irresistible arguments in its favour. This subject will occupy much of our attention in the history of the *second period*, till which, the establishment of the

* Mémoires de l'Acad. des Sciences, 1718, p. 245.

Newtonian philosophy on the continent cannot be said to have been accomplished.

In addition to these discoveries in physical astronomy, this period affords several in the descriptive parts of the science, of which, however, I can only mention one, as far too important to be passed over in the most general outline. It regards the apparent motion in the fixed stars, known by the name of the Aberration, and is the discovery of Dr Bradley, one of the most distinguished astronomers of whom England has to boast. Bradley and his friend Molyneux, in the end of the year 1725,* were occupied in searching for the parallax of the fixed stars by means of a zenith sector, constructed by Graham, the most skilful instrument-maker of that period. The sector was erected at Kew; it was of great radius, and furnished with a telescope twenty-four feet in length, with which they proposed to observe the transits of stars near the zenith, according to a method that was first suggested by Hooke, and pursued by him so far as to induce him to think that he had actually discovered the parallax of γ *Draconis*, the bright star in the head of the dragon, on which he made his observations. They began their observations of the transits of the same star on the 3d of December, when the distance from the zenith at which it passed was carefully

* Phil. Trans. Vol. XXXV. p. 697.

marked. By the observations of the subsequent days the star seemed to be moving to the south; and about the beginning of March, in the following year, it had got 20″ to the south, and was then nearly stationary. In the beginning of June it had come back to the same situation where it was first observed, and from thence it continued its motion northward till September, when it was about 20″ north of the point where it was first seen, its whole change of declination having amounted to 40″.

This motion occasioned a good deal of surprise to the two observers, as it lay the contrary way to what it would have done if it had proceeded from the parallax of the star. The repetition of the observations, however, confirmed their accuracy; and they were afterwards pursued by Dr Bradley, with another sector constructed also by Graham, of a less radius, but still of one sufficiently great to measure a star's zenith distance to half a second. It embraced a larger arch, and admitted of the observations being extended to stars that passed at a more considerable distance from the zenith.

Even with this addition the observations did not put Bradley in possession of the complete fact, as they only gave the motion of each star in declination, without giving information about what change might be produced in its right ascension.

Had the whole fact, that is, the motion in right

ascension as well as in declination been given from observation, it could not have been long before the cause was discovered. With such information, however, as Dr Bradley had, that discovery is certainly to be regarded as a great effort of sagacity. He has not told us the steps by which he was led to it; only we see that, by the method of exclusion, he had been careful to narrow the field of hypothesis, and had assured himself that the phenomenon was not produced by any nutation of the earth's axis; by any change in the direction of the plumb-line, or by refraction of any kind. All these causes being rejected, it occurred to him that the appearances might arise from the progressive motion of light combined with the motion of the earth in its orbit. He reasoned somewhat in this manner. If the earth were at rest, it is plain that a telescope, to admit a ray of light coming from a star to pass along its axis, must be directed to the star itself. But, if the earth, and, of course, the telescope be in motion, it must be inclined forward, so as to be in the diagonal of a parallelogram, the sides of which represent the motion of the earth, and the motion of light, or in the direction of those motions, and in the ratio of their velocities. It is with the telescope just as with the vane at the mast-head of a ship; when the ship is at anchor, the vane takes exactly the direction of the wind; when the ship is under weigh, it places itself in the diagonal of a parallelogram, of which one side represents the velocity of the ship, and the other the velocity of the wind. If, instead of the vane, we conceive a hollow tube, moveable in the same manner, the case will become more exactly parallel to that of the telescope. The tube will take such a position that the wind may blow through it without striking against the sides, and its axis will then be the diagonal of the parallelogram just referred to.

The telescope, therefore, through which a star is viewed, and by the axis of which its position is determined, must make an angle with the straight line drawn to the star, except when the earth moves directly upon the star, or directly from it. Hence it follows, that if the star be in the pole of the ecliptic, the telescope must be pointed forward, in the direction of the earth's motion, always by the same angle, so that the star would be seen out of its true place by that angle, and would appear to describe a circle round the pole of the ecliptic, the radius of which, subtended at the earth, an angle, of which the sine is to unity, as the velocity of the earth to the velocity of light. If the star be any where between the plane of the ecliptic and the pole, its apparent path will be an ellipse, the longer axis of which is the same with the diameter of the former circle, and the shorter equal to the same quantity, multiplied by the sine of the star's latitude. If the star be in the plane of the ecliptic, this shorter axis vanishes, and the apparent path of the star is a straight line, equal to the axis just mentioned.

Bradley saw that Roemer's observation concerning the time that light takes to go from the sun to the earth gave a ready expression for the velocity of light compared with that of the earth. The proportion, however, which he assumed as best suited to his observations was somewhat different; it was that of 10313 to 1, which made the radius of the circle of aberration 20″, and the transverse axis of the ellipse in every case, or the whole change of position, 40″. It was the shorter axis which Bradley had actually observed in the case of γ *Draconis*, that star being very near the solstitial colure, so that its changes of declination and of latitude are almost the same. In order to show the truth of his theory, he computed the aberration of different stars, and, on comparing the results with his observations, the coincidence appeared almost perfect, so that no doubt remained concerning the truth of the principle on which he had founded his calculations. He did not explain the rules themselves: Clairaut published the first investigation of these in the *Memoirs of the Academy of Sciences* for 1737. Simpson also gave a demonstration of them in his *Essays*, published in 1740.

It has been remarked, that the velocity of light, as assumed by Bradley, did not exactly agree with that which Roemer had assigned; supposing the total amount of the aberration $40\frac{1}{2}''$, it gave the time that light takes to come from the sun to the earth $8'\ 13''$; but it is proper to add, that since the time of this astronomer, the velocity of light deduced from the eclipses of Jupiter's satellites has been found exactly the same.

It is remarkable that the phenomenon thus discovered by Bradley and Molyneux, when in search of the parallax of the fixed stars, is in reality as convincing a proof of the earth's motion in its orbit, as the discovery of that parallax would have been. It seems, indeed, as satisfactory as any evidence that can be desired. One only regrets, in reflecting on this discovery, that the phenomenon of the aberration was not foreseen, and that, after being predicted from theory, it had been ascertained from observation. As the matter stands, however, the discovery both of the fact and the theory is highly creditable to its author.

In the imperfect outline which I have now sketched of one of the most interesting periods in the history of human knowledge, much has been omitted, and many great characters passed over, lost, as it were, in the splendour of the two great luminaries which marked this epocha. Newton and Leibnitz are so distinguished from the rest even of the scientific world, that we can only compare them with one another, though, in fact, no two intellectual characters, who both reached the highest degree of excellence, were ever more dissimilar.

For the variety of his genius, and the extent of his research, Leibnitz is perhaps altogether unrivalled. A lawyer, a historian, an antiquary, a poet, and a philologist,—a mathematician, a metaphysician, a theologian, and I will add a geologist, he has in all these characters produced works of great merit, and in some of them of the highest excellence. It is rare that original genius has so little of a peculiar direction, or is disposed to scatter its efforts over so wide a field. Though a man of great inventive powers, he occupied much of his time in works of mere labour and erudition, where there was nothing to invent, and not much of importance to discover. Of his inventive powers as a mathematician we have already spoken; as a metaphysician, his acuteness and depth are universally admitted; but metaphysics is a science in which there are few discoveries to be made, and the man who searches in it for novelty, is more likely to find what is imaginary than what is real. The notion of the *Monads*, those unextended units, or simple essences, of which, according to this philosopher, all things corporeal and spiritual, material or intellectual, are formed, will be readily allowed to have more in it of novelty than truth. The *pre-established harmony* between the body and the mind, by which two substances incapable of acting on one another, are so nicely adjusted from the beginning, that their movements for ever correspond, is a system of which no argument can do more than prove the possibility. And, amid all the talent and acuteness with which these doctrines are supported, it seems to argue some unsoundness of understanding, to have thought that they could ever find a place among the established principles of human knowledge.

Newton did not aim at so wide a range. Fortunately for himself and for the world, his genius was more determined to a particular point, and its efforts were more concentrated. Their direction was to the accurate sciences, and they soon proved equally inventive in the pure and in the mixed mathematics. Newton knew how to transfer the truths of abstract science to the study of things actually existing, and, by returning in the opposite direction, to enrich the former by ideas derived from the latter. In experimental and inductive investigation, he was as great as in the pure mathematics, and his discoveries as distinguished in the one as in the other. In this double claim to renown, Newton stands yet unrivalled; and though, in the pure mathematics, equals may perhaps be found, no one, I believe, will come forward as his rival both in that science and in the philosophy of nature. His caution in adopting general principles; his dislike to what was vague or obscure; his rejection of all theories from which precise conclusions cannot be deduced; and his readiness to relinquish those that depart in any degree from the truth, are, throughout, the characters of his philosophy, and distinguish it very essentially from the philosophy of Leibnitz. The characters now enumerated are most of them negative, but without the principles on which they are founded, invention can hardly be kept in the right course. The German philosopher was not furnished with them in the same degree as the English, and hence his great talents have run very frequently to waste.

It may be doubted also, whether Leibnitz's great metaphysical acuteness did not sometimes mislead him in the study of nature, by inclining him to those reasonings which proceed, or affect to proceed, continually from the cause to the effect. The attributes of the Deity were the axioms of his philosophy; and he did not reflect that this foundation, excellent in itself, lies much too deep for a structure that is to be raised by so feeble an architect as man; or, that an argument, which sets out with the most profound respect to the Supreme Being, usually terminates in the most unwarrantable presumption. His reasonings from first causes are always ingenious; but nothing can prevent the substitution of such causes for those that are physical and efficient, from being one of the worst and most fatal errors in philosophy.

As an interpreter of nature, therefore, Leibnitz stands in no comparison with Newton. His general views in physics were vague and unsatisfactory; he had no great value for inductive reasoning; it was not the way of arriving at truth which he was accustomed to take; and hence, to the greatest physical discovery of that age, and that which was established by the most ample induction, the existence of gravity as a *fact* in which all bodies agree, he was always incredulous, because no proof of it, *à priori*, could be given.

As to who benefited human knowledge the most, no question, therefore, can arise; and if genius is to be weighed in this balance, it is evident which scale must preponderate. Except in the pure mathematics, Leibnitz, with all his talents, made no material or permanent addition to the sciences. Newton, to equal inventions in mathematics, added the greatest discoveries in the philosophy of nature; and, in passing through his hands, Mechanics, Optics, and Astronomy, were not merely improved, but renovated. No one ever left knowledge in a state so different from that in which he found it. Men were instructed not only in new truths, but in new methods of discovering truth; they were made acquainted with the great principle which connects together the most distant regions of space, as well as the most remote periods of duration; and which was to lead to future discoveries, far beyond what the wisest or most sanguine could anticipate.

END OF VOLUME SECOND.

Printed by George Ramsay & Co.
Edinburgh, 1821.

Printed in the United States
By Bookmasters